99 Topics in Current Chemistry

Fortschritte der Chemischen Forschung

Managing Editor: F. L. Boschke

Cosmo-
and Geochemistry

With Contributions by
E. Anders, R. Hayatsu, R. Lüst,
V. Marchig, G. Winnewisser

With 39 Figures and 18 Tables

Springer-Verlag Berlin Heidelberg GmbH 1981

This series presents critical reviews of the present position and future trends in modern chemical research. It is addressed to all research and industrial chemists who wish to keep abreast of advances in their subject.

As a rule, contributions are specially commissioned. The editors and publishers will, however, always be pleased to receive suggestions and supplementary information. Papers are accepted for "Topics in Current Chemistry" in English.

ISBN 978-3-662-15364-2   ISBN 978-3-540-38806-7 (eBook)

DOI 10.1007/978-3-540-38806-7

Library of Congress Cataloging in Publication Data.
Main entry under title: Cosmo- and geochemistry.
(Topics in current chemistry — Fortschritte der Chemischen Forschung; 99)
Includes bibliographies and index.
1. Cosmochemistry —— Addresses, essays, lectures.
2. Geochemistry —— Adresses, essays, lectures. I. Anders, Edward, 1926 —
II. Series: Topics in current chemistry; 99.
QD1.F58 vol. 99 [QB450] 540s [523.02] 81–9330 AACR2

Originally published by Springer-Verlag Berlin Heidelberg New York in 1981.

Softcover reprint of the hardcover 1st edition 1981

Table of Contents

Organic Compounds in Meteorites and their Origins

Ryoichi Hayatsu[1] and Edward Anders[2]

[1] Chemistry Division, Argonne National Laboratory, Argonne, IL 60439, USA
[2] Enrico Fermi Institute and Dept. of Chemistry, University of Chicago, Chicago, IL 60637, USA

Table of Contents

Abstract

C1 and C2 carbonaceous chondrites contain several percent of organic matter, mainly as a bridged aromatic polymer containing COOH, OH, and CO groups, as well as heterocyclic rings containing N, O, and S. The remaining 5–30 % includes the following compound classes, either present initially or generated by solvolysis: alkanes (mainly normal), alkenes, arenes, alicyclics, alcohols, aliphatic carboxylic acids, purines, pyrimidines, and other basic N-compounds, amino acids, porphyrin-like pigments, carbynes, etc.

On the basis of laboratory experiments, it appears that these compounds formed in the solar nebula, by catalytic (Fischer-Tropsch) reactions of CO, H_2, and NH_3 at 360–400 K, $\sim 10^{-5}$ atm. The onset of these reactions was triggered by the formation of suitable catalysts (magnetite, hydrated silicates) at these temperatures. Such reactions may be a source of prebiotic carbon compounds on the inner planets, and interstellar molecules.

1 Introduction

Type 1 carbonaceous chondrites (= C1) are the most primitive samples of matter known to man. They contain all stable elements in solar proportions, with the exception of the noble gases and the highly volatile major elements H, C, N, O. It is generally agreed that they are low-temperature condensates from the solar nebula (Whipple, 1964; Anders, 1964, 1971; Larimer and Anders, 1967; Levin, 1969; Grossman, 1972; J. S. Lewis, 1972; Grossman and Larimer, 1974). But they also contain small amounts of alien, presolar matter (R. N. Clayton, 1978; R. S. Lewis et al., 1979), and were altered by liquid water in the meteorite parent bodies (DuFresne and Anders, 1962; Nagy et al., 1963; Bunch and Chang, 1980). Thus at least some properties of C1's actually pre-date or post-date the nebular stage; for maximalist views on this question, see Cameron (1973), D. D. Clayton (1978); Kerridge et al. (1979), and Bunch and Chang (1980).

C1 chondrites contain 6% of their cosmic complement of carbon, mainly in the form of organic matter. The intense controversy that once surrounded the origin of this organic matter has subsided. Most authors now agree that this material represents primitive prebiotic matter, not vestiges of extraterrestrial life. The principal questions remaining are what abiotic processes formed the organic matter, and to what extent these processes took place in locales other than the solar nebula: interstellar clouds or meteorite parent bodies.

We can approach the problem in three stages. First, we try to reconstruct the physical conditions during condensation (temperature, pressure, time) from the clues contained in the inorganic matrix of the meteorite. Next, we determine the condensation behavior of carbon under these conditions, on the basis of thermodynamic calculations. Finally, we perform model experiments on the condensation of carbon, and compare the compounds synthesized with those actually found in meteorites.

The present paper is an extensively updated version of an earlier review (Anders et al., 1973). It covers the literature through mid-1980.

2 Formation Conditions of Carbonaceous Chondrites

Temperature. Six different "cosmothermometers", based on kinetic or equilibrium isotope fractionations, or on chemical reactions, consistently give temperatures near 360 K for C1 chondrites (Table 1). These values represent the temperatures at which isotopic or chemical equilibria were frozen in, owing to sluggish reaction rates or physical isolation of the meteoritic dust from the ambient gas (by accretion to larger bodies). These temperatures are lower than those for C2 chondrites, 380 to 400 K (Onuma et al., 1972, 1974; Lancet, 1972) or ordinary chondrites and achondrites, ~450 to ~500 K (Grossman and Larimer, 1974).

Pressure. There are no reliable pressure estimates for C1's. Tentative values of ~10^{-5} atm were obtained for ordinary chondrites from their Bi, T1, and In contents (Larimer, 1973). An upper limit of <2×10^{-3} atm has been inferred by Grossman and Clark (1973) for C3 chondrites from the lack of evidence for gas-to-

Table 1. Formation Temperatures of C1 Chondrites

Thermometer	T (K)	P (atm)	Reference
O^{18}/O^{16} $CaMg(CO_3)_2$-H_2O	360 ± 5	Independent	Onuma et al. (1972)
O^{18}/O^{16} serpentine-H_2O	360 ± 15	Independent	Onuma et al. (1972)
C^{13}/C^{12} $CaMg(CO_3)_2$-polymer	357 ± 21	Independent	Lancet (1972)
$3 Fe + 4 H_2O \leftrightarrows Fe_3O_4 + 4 H_2$	≤ 400	Independent	Urey (1952b)
Olivine + $H_2O \leftrightarrows$ serpentine	~350	10^{-4}	Larimer and Anders (1967)
Tl abundance	< 425	10^{-5}	Larimer (1973)

liquid condensation. These values are consistent with theoretical models of the solar nebula, which predict $P \simeq 10^{-5}$ atm, $T = 300$–400 K for the asteroid belt (Cameron and Pine, 1973). More recent work on the evolution of the nebula suggests, however, that a major part of planetary matter was cycled through gaseous protoplanets (Cameron, 1978), and hence may have experienced higher pressures, to ≥ 1 atm. It is not clear, though, that C1 chondrites ever were located in such protoplanets, and so a pressure of ~ 10^{-5} atm may still be appropriate.

Time. An age determination based on extinct 16 Myr I^{129} gave an age difference of only 0.2 ± 0.2 Myr for magnetites from a C1 and a C2 chondrite (Lewis and Anders, 1975), suggesting that the entire condensation stage of the solar nebula was of that order. However, it now appears that I^{129} "ages" reflect some combination of true age differences and variations in the initial I^{129}/I^{127} ratio (Podosek, 1978; Jordan et al., 1980; Crabb et al., 1980). The best estimate of the time scale comes from theoretical models, showing consistently that the lifetime of the nebula cannot have been more than a few times 10^5 yr (Cameron, 1978).

3 Behavior of Carbon in a Solar Gas

3.1 Equilibrium

With the physical conditions thus defined, it is instructive to consider what happens to carbon in a cooling (or contracting) solar gas (Fig. 1). CO is the stable form at high temperatures or low pressures, but becomes less stable on cooling or compression and should transform to CH_4 below 600 K at 10^{-5} atm, as shown by the solid lines in the right-hand portion of Fig. 1. However, CH_4 has a condensation temperature of less than 100 K, and if this reaction had gone as written, there should be no carbon and no life anywhere in the inner solar system. Since there is at least some evidence to the contrary (C. Sagan, private communication), events must have taken a different course.

Urey (1953) first noted this paradox in a classic paper. After finding that *elemental* C could condense only at $T < 400$ K, $P < 10^{-10}$ atm (a thorough reinvestigation by J. S. Lewis et al., 1979, gives somewhat less restrictive limits), he suggested that the CO-CH_4 transformation might not have gone smoothly "in the absence of man-devised catalysts", but might instead have "proceed[ed] through graphite or

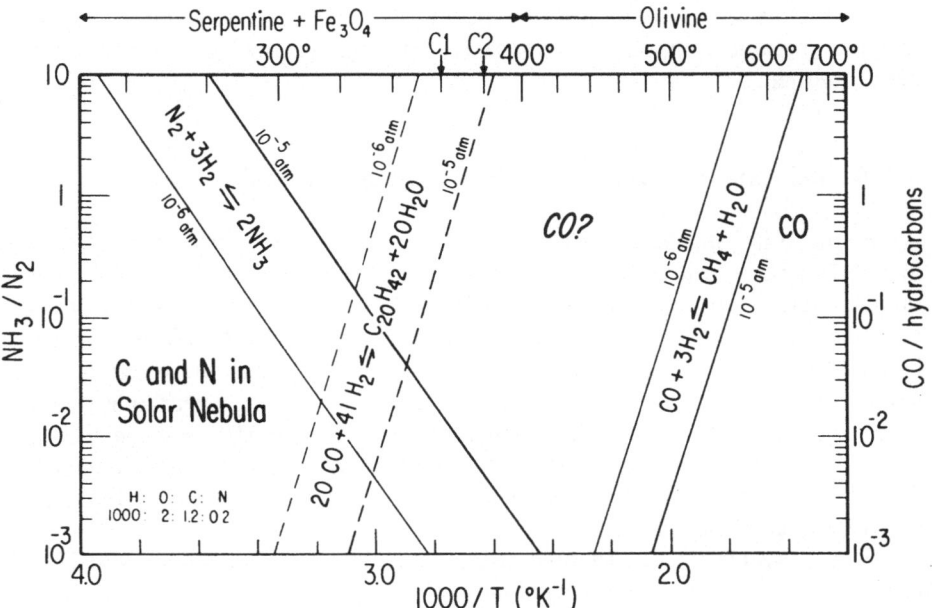

Fig. 1. If equilibrium is maintained on cooling, CO will be converted largely to CH_4 (solid lines) before metastable formation of more complex hydrocarbons by the Fischer-Tropsch reaction becomes possible (dashed lines). However, the reaction is very slow in the absence of catalysts, and may not have begun until about 400 K, when catalysts such as serpentine and magnetite became available through the hydration of olivine. Thus CO may have persisted metastably between 600 and 400 K

complex tarry carbon compounds as intermediates. Such compounds have long been known to be constituents of the carbonaceous chondritic meteorites. It seems most probable to the writer that such compounds constitute the non-volatile carbon compounds which supplied the carbon to the earth and meteorites while other gases, the hydrogen and inert gases, were lost from the region of the forming earth".

This idea soon fell into oblivion; ironically, due to the success of another of Urey's ideas. A major paradox facing all theories on the origin of life on Earth was that the CO_2 in the Earth's atmosphere would not *spontaneously* transform to organic compounds, all of which are thermodynamically less stable. Calvin (1969) showed in 1950 that the thermodynamic problem could be overcome by supplying energy in the form of ionizing radiation. But the crucial advance came two years later when Urey (1952a), following Oparin (1927), realized that the primitive atmosphere must have been highly reduced (CH_4 rather than CO_2, along with NH_3, H_2O, and H_2). He suggested that ultraviolet light or electric discharges might have converted the molecules in the atmosphere to excited states, free radicals, or other reactive species, which would then transform spontaneously into a variety of organic compounds, depending on the available kinetic pathways.

Urey's suggestion was soon confirmed by Miller (1953) in a historic experiment, in which amino acids were produced from CH_4, H_2O, and NH_3 by an electric spark discharge. [The analogous synthesis with UV light was not achieved with high efficiency until much later (Sagan and Khare, 1971)]. A new scientific discipline, abiotic synthesis, was born. In the ensuing excitement, Urey's other suggestion (complex organic compounds from partial hydrogenation of CO) seems to have been forgotten.

3.2 Model Experiments: The Fischer-Tropsch Synthesis

We began to investigate this neglected idea in 1964, when it first became apparent that Miller-Urey reactions could not account for certain features of meteorite organic matter. Our approach was to see how CO and H_2 behaved in the presence of some *natural* catalysts expected in the solar nebula: nickel-iron, magnetite, hydrated silicates. We found (Studier et al., 1968, 1972) that the reaction indeed tends to stop at intermediate stages of hydrogenation, giving metastable products of $H/C \sim 2$ (e.g. $C_{20}H_{42}$) rather than stable methane of $H/C = 4$. In fact, this process has been used industrially for the production of gasoline: it is the Fischer-Tropsch synthesis, nominally dating from 1923, but first carried out by Döbereiner in 1817 (Bauer, 1980).

It is not feasible to conduct such model experiments at the pressure and H_2/CO ratio of the solar nebula — the total amount of carbon in a 1-liter vessel would be only $\sim 10^{-8}$ grams. Accordingly, we used higher pressures (0.1–10 atm), and lower H_2/CO ratios (generally 1, sometimes as high as 120). We shall show later on that these results are nonetheless applicable to the solar nebula.

As a safeguard against contamination, we used deuterium rather than light hydrogen in our syntheses, and identified reaction products by mass spectrometry (= MS). Perdeuterated compounds give peaks only at even mass numbers (except for very minor peaks due to the rare isotope C^{13}). Contaminants, on the other hand, give prominent peaks at both even and odd mass numbers. The analytical methods, described by Studier et al. (1978), evolved over the years and included GCMS; vacuum distillation or pyrolysis-MS; variable-temperature, solid probe MS; high-resolution MS; liquid, paper, or thin-layer chromatography with or without MS; and IR, UV, or NMR spectroscopy.

The reactions were always carried out unter static conditions in a closed vessel, not in a flow system as in the industrial synthesis. We also broadened the range of experimental conditions beyond those of the classical Fischer-Tropsch synthesis. Reaction times ranged from a fraction of an hour to a few months, and the temperature was sometimes raised briefly from 150–250 °C to 500–700 °C. Such thermal pulses were intended to simulate short-term heating events in the nebula, such a collisions, shock waves, or the chondrule-forming process, whatever its nature. When nitrogen compounds were sought we added ND_3 to the reactant gases, and occasionally used other types of catalyst, such as montmorillonite clay, Al_2O_3, or SiO_2. For want of a better term, we shall call this class of reactions "Fischer-Tropsch-type" (= FTT).

3.3 Non-Spontaneous vs Spontaneous Reactions

The basic dilemma of abiotic synthesis is that the organic compounds sought are thermodynamically unstable with respect to the low-T equilibrium forms of C, N,

O, and H: CH_4, NH_3, H_2O, and H_2. The Miller-Urey reaction circumvents this problem by infusing energy (UV, electric discharges), thus producing free radicals or unstable molecules that can react further to yield organic compounds. The FTT reaction, on the other hand, uses CO, the high-T/low-P equilibrium form of carbon in a solar gas, whose reactions to form organic compounds are exoergic[1] and spontaneous, requiring no external energy source.

It appears that conditions in the solar nebula were appropriate for the FTT but not the Miller-Urey reaction. Kinetic calculations (Lewis and Prinn, 1980) as well as observations on comets (Delsemme, 1977) show that CO and CO_2, not CH_4, were the principal forms of carbon. And the dust-laden solar nebula was opaque to UV, precluding any photochemical reactions. It seems best, however, to approach the problem empirically, by examining the meteoritic organic compounds themselves for clues to their formation. We shall review these compounds class by class, looking for the signatures of the FTT or Miller-Urey reactions.

4 Organic Compounds in Meteorites

There is little doubt that the bulk of the organic matter in meteorites is indigenous, judging from isotopic measurements on C, H, and S (Briggs, 1963; Smith and Kaplan, 1970; Kolodny et al., 1980; Robert et al., 1980). Such proof is not available for individual compounds, however, and one must therefore be on the alert for contamination. We shall briefly review the principal compound classes, commenting on their authenticity and origin. Owing to space limitations, we can only give a concise summary of the most pertinent data. The interested reader may therefore wish to consult the original sources, as well as the principal reviews (Hayes, 1967; Vdovykin, 1967, 1979; Oró, 1972; Stephen-Sherwood and Oró, 1973; Kvenvolden, 1974; and Nagy, 1975).

4.1 Hydrocarbons

Heavy Alkanes. Although there is some variation among individual meteorites and FTT synthesis runs, the same few compounds dominate in both (Meinschein, 1963; Studier et al., 1968, 1972; Studier and Hayatsu, 1968; Gelpi et al., 1970;

---

[1] Such exoergic reactions with reactive molecules have often been called "thermal", with the implication that "thermal energy" somehow supplies the driving force. However, this term is misleading. The driving force is the inherent instability of these molecules at low temperatures, which permits them to transform to organic compounds, with a net decrease in free energy of the system. The principal way in which "thermal energy" enters is in speeding up reaction rates. Thus such reactions should be called "spontaneous" rather than "thermal".

Actually, much of the experimental work on chemical evolution (cf. Lemmon, 1970) utilizes such unstable compounds, e.g. HCN, HCHO, $HC \equiv CCN$, H_2NCN, etc., on the grounds that they can be made by Miller-Urey reactions. But they can also be made by spontaneous reactions of CO, NH_3, and H_2 (Anders et al., 1974). Hence this class of reactions provides some common ground between the two main types of abiotic synthesis.

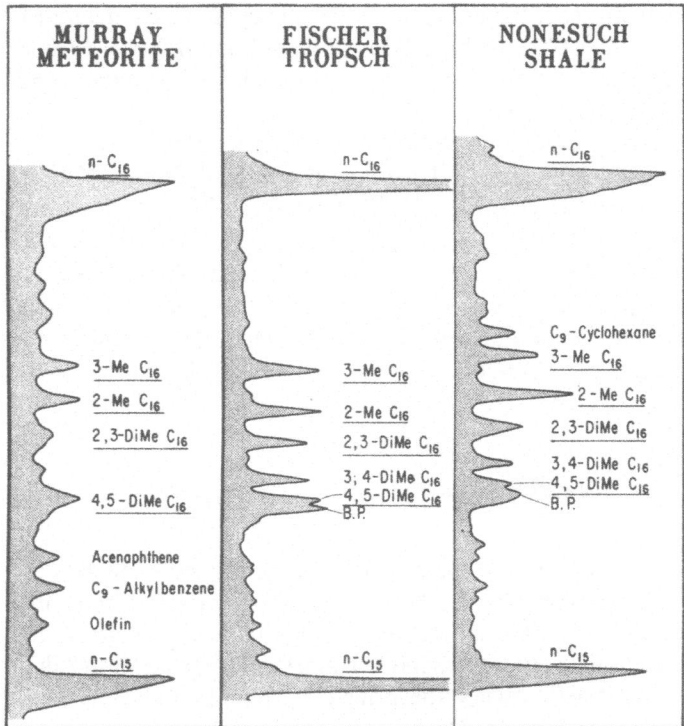

Fig. 2. Gas chromatogram of hydrocarbons in the range C_{15} to C_{16} (Studier et al., 1972). Only 6 of the $\sim 10^4$ isomeric hydrocarbons with 16 C atoms are present in appreciable abundance; 5 of them (underlined) are common to all three samples. The sample of the Precambrian Nonesuch shale (courtesy W. G. Meinschein) is a pure aliphatic fraction from which aromatic hydrocarbons had been removed by silica gel chromatography; it will be discussed in Sec. 8.2. Abbreviations: Me, methyl; B.P., branched paraffin

Nooner and Oró, 1967). Normal (straight-chain) alkanes are most prominent, followed by methyl and dimethyl alkanes and structurally similar, slightly branched alkenes (Fig. 2).

This resemblance is highly significant if one considers that 10,359 structural isomers exist for saturated hydrocarbons with 16 C atoms (Lederberg, 1972). Apparently the meteoritic hydrocarbons were made by FTT reactions, or some other process of the same extraordinary selectivity. The Miller-Urey reaction, incidentally, shows no such selectivity. Gas chromatograms of hydrocarbons made by electric discharges in methane show no structure whatsoever in the region around C_{16} (Ponnamperuma et al., 1969). Apparently all 10^4 possible isomers are made in comparable yield, as expected for random recombination of free radicals.

The indigenous nature of the alkanes in at least the Murray, Murchison, and Orgueil meteorites (Studier et al., 1968, 1972) is supported by 4 lines of evidence, to be discussed later in this paper: absence of the isoprenoids pristane and phytane that occur in nearly all terrestrial hydrocarbon samples; characteristic light hydrocarbon pattern; low abundance of alkanes in the C3 chondrite Allende (Levy et al., 1970; Studier et al., 1972) which is thermally metamorphosed and hence may be regarded as a blank; and carbon isotope data (Sect. 5.1).

Aromatic Hydrocarbons and Light Alkanes. Carbonaceous chondrites contain a wide range of aromatic hydrocarbons, from benzene through alkylbenzenes and -naphthalenes to polynuclear hydrocarbons of up to six fused benzene rings (see the reviews cited above). At higher carbon numbers, aromatics tend to be less abundant than normal alkanes, but below about C_{11}, the reverse is true (Fig. 3, top). In fact, virtually no normal alkanes at all are found between C_2 and C_8, their place having been taken largely by benzene, toluene, xylene, and various alkenes or branched alkanes, notably butene (Studier et al., 1965b, 1968, 1972; Hayes and Biemann, 1968; Levy et al., 1973).[2]

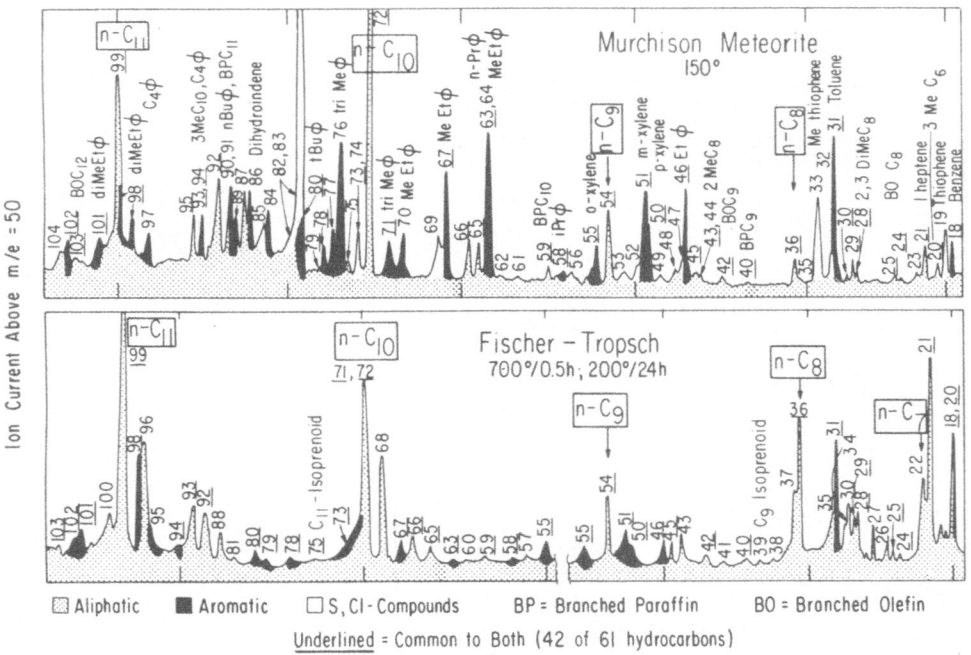

Fig. 3. Hydrocarbons from Murchison meteorite and Fischer-Tropsch synthesis. BP = branched paraffin; BO = branched olefin; ϕ = phenyl radical. For additional peak identifications, see Studier et al. (1972). Of the 61 hydrocarbons in the meteorite, 42 (underlined) are also present in the Fischer-Tropsch sample, though often not in comparable amount

[2] Contrary to these findings, Belsky and Kaplan (1970) have reported substantial amounts of C_2 to C_6 alkanes and alkenes, but only small amounts of benzene and toluene. However, their work is at variance with that of all other authors cited above, and has been criticized in detail by Studier et al. (1972).

A pattern of this sort does not form directly in the primary Fischer-Tropsch reaction. It does, however, develop when a primary Fischer-Tropsch mixture remains in contact with the catalyst, for a day or so at 350–400 °C (Fig. 3, bottom), or longer times at lower temperatures (Studier et al., 1968, 1972; Galwey, 1972). Under such conditions, a metastable equilibrium is approached, with methane and aromatic hydrocarbons forming at the expense of ethane and heavier alkanes (Dayhoff et al., 1964; Eck et al., 1966). The kinetics and mechanism of such aromatization on the catalyst surface has been discussed by Galwey (1972). Of the 61 hydrocarbons in the meteorite, 42 (underlined) are also seen in the synthetic sample, though often not in the same amount. It remains to be seen whether the match can be made more quantitative by changes in the reheating conditions.

When the heating is prolonged or carried out at higher temperatures, polynuclear aromatic hydrocarbons with up to 7 rings form (Fig. 4; Studier et al., 1965a, 1968; Oró and Han, 1966; Friedman et al., 1970). This reaction, using CH_4 as the starting material, was discovered by Berthelot over a century ago.

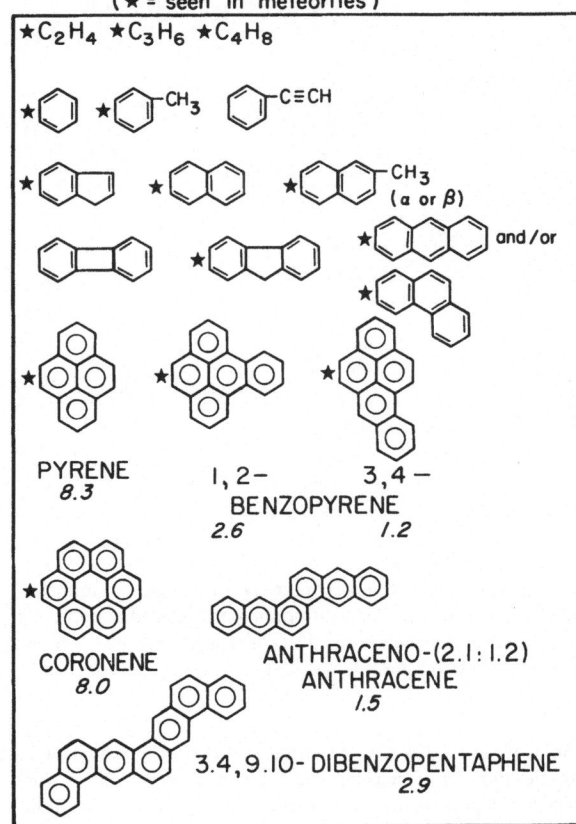

Hydrocarbons from Pyrolysis of CH_4 or CH_4-CO-CO_2 mixtures at 900°C, 2-39h.
(★ = seen in meteorites)

★C_2H_4 ★C_3H_6 ★C_4H_8

PYRENE
8.3

1, 2–
BENZOPYRENE
2.6

3,4 –
1.2

CORONENE
8.0

ANTHRACENO-(2.1:1.2)
ANTHRACENE
1.5

3.4,9.10- DIBENZOPENTAPHENE
2.9

Fig. 4. Hydrocarbons produced from CH_4 or CH_4—CO—CO_2 mixtures at 900 °C (Studier et al., 1965b, 1968). Italicized numbers are percent yields, relative to 100% total carbon. Many of these compounds are also present in meteorites (Vdovykin, 1967; Pering and Ponnamperuma, 1971; Studier et al., 1972)

Opportunities for such secondary reactions certainly existed in the history of meteorites. Temperatures in the nebula (360–400 K, Table 1) may alone have been high enough for secondary reactions in the time available, $\sim 10^4$–10^5 yr. Kinetic studies of a similar reaction (formation of benzene from alcohols, amines, or fatty acids on Fe_2O_3 or iron-rich peat catalysts; Galwey, 1972) indicate a benzene formation rate of 5×10^{16} molecules g^{-1} yr^{-1} at 360 K. At this rate it would take only 5000 years to transform all the meteoritic carbon to benzene. Further opportunities were provided by brief thermal pulses during chondrule formation, impact, or transient shocks. Of course, any high-temperature episodes must have happened early or on a local scale, to permit survival of other, more temperature-sensitive compounds.

Isoprenoid Alkanes. Early work suggested (Nooner and Oró, 1967; Gelpi and Oró, 1970a) that nearly all carbonaceous chondrites contain the isoprenoid alkanes pristane and phytane (2-,6-,10-,14-tetramethyl-pentadecane and -hexadecane). These two hydrocarbons, which may formally be regarded as tetramers of isoprene, $CH_2 : C(CH_3)CH : CH_2$, serve as biological markers on Earth, being derived mainly

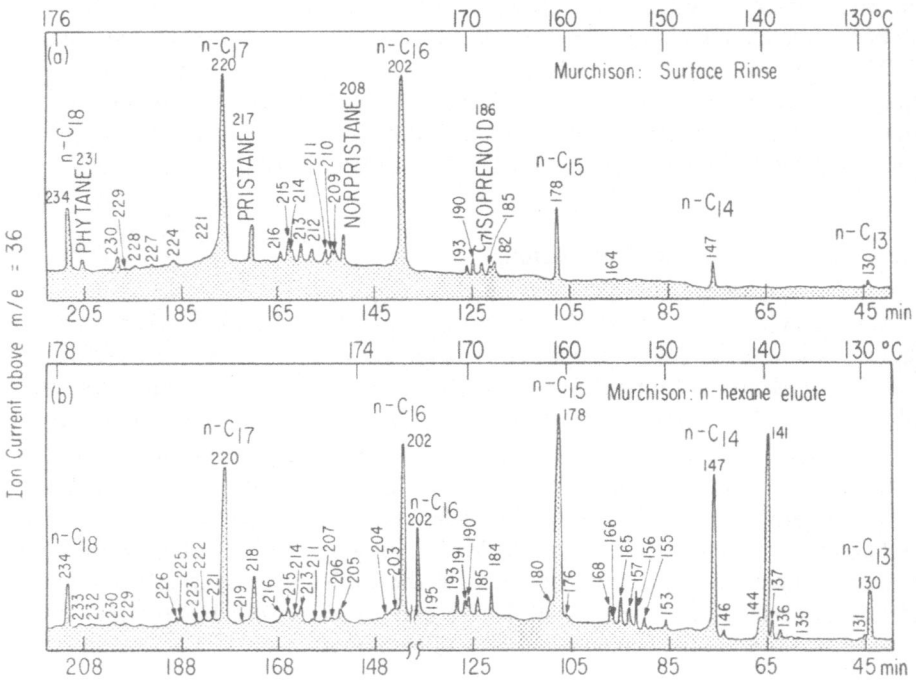

Fig. 5a. A surface rinse of Murchison (benzene-methanol and *n*-hexane at room temperature) contains 4 isoprenoid hydrocarbons (capitalized). For peak identifications, see Studier et al. (1972). **b** Benzene-methanol extract of the same sample after grinding still yields a variety of alkanes and alkenes, but no isoprenoids. Peak 218 preceding n-C_{17} is not pristane but an alkene. Apparently the heavier isoprenoids are surficial contaminants of terrestrial origin

from the phytol side chain of the chlorophyll molecule. Their presence in meteorites thus suggested either extraterrestrial life or an abiotic process that strongly favored isoprenoids over other types of branched hydrocarbons (Calvin, 1969; McCarthy and Calvin, 1967).

It seems, however, that these results reflect terrestrial contamination. Studier et al. (1968) found no tetrameric isoprenoids in Orgueil and Murray, only small amounts of dimeric isoprenoids from C_9 to C_{14}. The latter compounds can, however, be produced in FTT syntheses (Studier et al., 1968, 1972; Pichler et al., 1968); in fact, they are more prominent in synthetic material than in meteorites (Fig. 3). A sample of the Murchison meteorite gave substantial amounts of tetrameric and trimeric isoprenoids, but they were found only in a surface rinse of the stone, not in an interior extract (Fig. 5). Yet the other alkanes persisted in the interior, and since most petroleums and other terrestrial alkane mixtures contain isoprenoids (Calvin, 1969; Gelpi and Oró, 1970a), this would seem to strengthen the case for an extraterrestrial origin of these alkanes. Gelpi and Oró (1970b) cast further doubts on the earlier identifications by showing that pristane, phytane and other heavy isoprenoids are prominent in house dust and even in hydrocarbon extracts from FeS nodules in iron meteorites.

Alicyclic Hydrocarbons. Low-resolution GS or GCMS studies have suggested the presence of alicyclics, such as cycloalkanes (Hayes, 1967) or bicyclics with aliphatic side chains (Kvenvolden et al., 1970; Oró et al., 1971). High-resolution GCMS, however, confirmed only cyclopentene (Studier et al., 1972). An earlier study detected a cyclic terpene, Δ^8 and/or $\Delta^{4(8)}$ menthene, but it may have been a contaminant (Studier et al., 1968). In any event, the Fischer-Tropsch synthesis readily produces alicyclic hydrocarbons (Weitkamp et al., 1953).

4.2 Oxygen-Containing Compounds

Alcohols and Carbonyl Compounds. These compounds have been somewhat neglected. Though Studier et al. (1965b) reported evidence for alcohols in Murray and Breger et al. (1972) identified formaldehyde in Allende, the first systematic study was done only in 1976, when Jungclaus et al. identified alcohols (C_1-C_4), aldehydes (C_2-C_4) and ketones (C_3-C_5) in water extracts of the C2 chondrites Murchison and Murray. Essentially the same compounds were made in an FTT synthesis on a mixed, FeNi-clay catalyst (Anders et al., 1974). They also are produced in the industrial Fischer-Tropsch synthesis (Cain et al., 1953; Asinger, 1968). Thus far only a few of these compounds have been identified in the Miller-Urey synthesis (methanol, ethanol, acetone, and acetaldehyde; Khare and Sagan, 1973), but in view of the limited effort, no negative conclusions should be drawn.

Aliphatic Carboxylic Acids. Nagy and Bitz (1963) reported fatty acids from C_{14} to C_{28} in Orgueil. This work was substantially confirmed by Hayatsu (1965) and Smith and Kaplan (1970). The latter authors found 3 to 91 ppm fatty acids from C_{12} to C_{20} in 7 carbonaceous chondrites. Smith and Kaplan believe that these acids are largely or entirely terrestrial contaminants, because unstable, unsaturated acids comprise about 30% of the total, and C_{16} and C_{18} acids are predominant. The case for contamination was strengthened by Yuen and Kvenvolden (1973), who found C_{14}, C_{16}, and C_{18} acids only in surface, but not interior, samples of Murchison. However, the interior samples contained a number of short-chain acids, of which 10 were definitely identified (C_1–C_8) and five others (C_5–C_7, all branched) were tentatively identified.

Extending this work, Lawless and Yuen (1979) made a quantitative study of monocarboxylic acids in Murchison. Eleven acids, ranging from C_2 to C_8, were identified. Of these, propanoic acid was the most abundant (1.83 µmol/g), exceeding the concentration of the most abundant amino acid, glycine, by a factor of 58.

Lawless and Yuen conclude that the Murchison fatty acids cannot have formed by the FT reaction, since the ratio normal/*total* branched is *less* than 1 (0.57–0.86 between C_4 and C_6). But this conclusion is based on two misconceptions. First, the denominator of this ratio includes the sum of *all* branched isomers; the more pertinent ratio for *single* isomers is considerably higher, e.g. 7 for C_6, still implying a strong preference for the normal isomer. Second, as we shall discuss in Sec. 8.2, the isomer distribution in the FT reaction depends on a single parameter, representing the relative probabilities of chain growth and chain branching (Friedel and Sharkey, 1963). This parameter varies greatly with temperature, time, catalyst, and H_2/CO ratio. For example, the normal/branched ratio for alkanes ranges from 52.6 at 195 °C to 1.2 at 370 °C, with iron meteorite and iron ore catalysts, respectively (Nooner et al., 1976). For fatty acids, the ratios reported are less variable (1.8 to 3.0 between C_6 and C_{19}), but so were the reaction conditions (Nooner and Oró, 1979). All were determined at a single temperature, 400 °C, and with a single type of catalyst, nickel-iron/alkali carbonate, rather than the more appropriate clays.

The Miller-Urey reaction fails qualitatively rather than quantitatively. It produces *no* detectable normal acids above C_6, even in the presence of an alkaline aqueous phase that is known to favor growth of linear chains by formation of a monolayer (Allen and Ponnamperuma, 1967). Given the fundamentally random nature of the Miller-Urey reaction, there is little hope that it will ever achieve the needed selectivity for normal isomers.

Seven α-hydroxycarboxylic acids (C_2 to C_5) and 17 aliphatic dicarboxylic acids (C_2 to C_9) have been found in Murchison (Peltzer et al., 1978; Lawless et al., 1974). Eight of the latter also were produced in a Miller-Urey synthesis (Zeitman et al., 1974). None have been looked for in the FTT synthesis thus far.

4.3 Purines, Pyrimidines, and Other Basic N-Compounds

Adenine, guanine, guanylurea, and several *s*-triazines and ureas have been detected in HCl-extracts of Orgueil and Murchison (Hayatsu, 1964; Hayatsu et al., 1968, 1975). The first 3 were confirmed by Stoks and Schwartz (1981), but the *s*-triazines were not; they may have formed from guanylurea in the isolation and identification procedure. Other compounds detected are xanthine, hypoxanthine, and uracil (van der Velden and Schwartz, 1977; Stoks and Schwartz, 1979). A report of 4-hydroxy-pyrimidine and several related compounds (Folsome et al., 1973) was not confirmed (Hayatsu et al., 1975; van der Velden and Schwartz, 1977); these compounds, which were identified as trimethyl silyl derivatives, may be artifacts formed during the silylation procedure.

Basic nitrogen compounds form in FTT reactions in the presence of NH_3 (Fig. 6). In the early experiments, the temperature was briefly raised to 500–700 °C (Hayatsu et al., 1968, 1972; Yang and Oró, 1971), but this may not be necessary, at least with a montmorillonite catalyst (Anders et al., 1974). The reaction mechanism is obscure, but probably involves reactive intermediates such as HCN, nitriles, or acetylenes (Hayatsu et al., 1968, 1972; Anders et al., 1974). It is not clear whether a liquid water phase, as generally formed in the cold neck of the vessel, is essential. A detailed, systematic study of this reaction would be very desirable, to see what conditions are required, and whether they are realistic for the solar nebula.

The Miller-Urey reaction has been notably less successful in producing N-heterocyclics. Only adenine has thus far been made, by electron irradiation of CH_4, NH_3, H_2O, and H_2 (Ponnamperuma et al., 1963). The yield was only 0.01 %. Better success was achieved by reactions involving unstable reactants, such as HCN (Lemmon, 1970), but these reactions, being spontaneous, actually are related no less to FTT than to Miller-Urey reactions (Sec. 3.3).

Hayatsu et al. (1975) suggested the presence of aliphatic amines in Murchison, but did not identify individual compounds. Jungclaus et al. (1976) found that at least 10 aliphatic amines (C_1–C_4), including 8 primary amines, were present in Murchison extracts. Several aliphatic amines have been seen in FTT syntheses (Kölbel and Trapper, 1966; Yoshino et al., 1971; Anders et al., 1974), but thus far not in the Miller-Urey synthesis.

Nucleotide Bases Made by Fischer-Tropsch-Type Synthesis

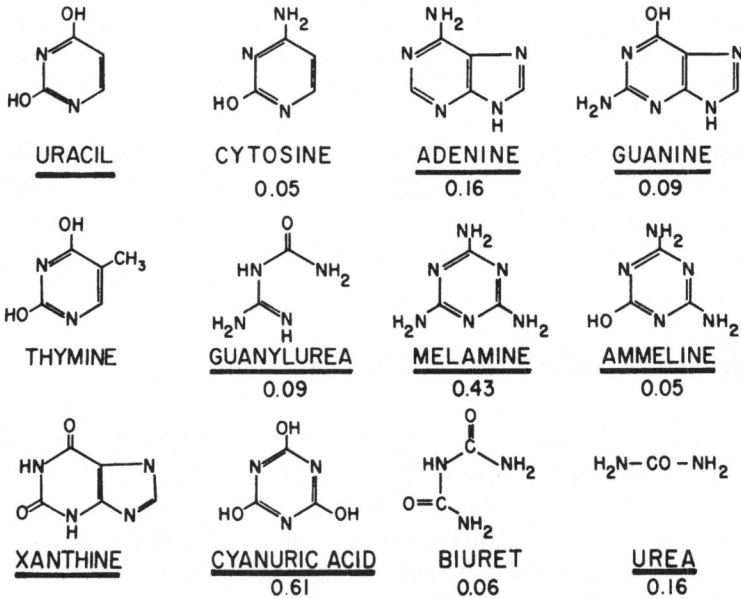

Fig. 6. Underlined compounds have been identified in meteorites; dashed lines represent tentative identifications. Numbers indicate percent yields in FTT synthesis (Hayatsu et al., 1968, 1972)

4.4 Amino Acids

Early reports of amino acids in meteorites were discredited by the discovery that fingerprints give an equally rich spectrum of amino acids, in comparable quantities (Oró and Skewes, 1965). However, Kvenvolden et al. (1970, 1971) showed by an elegant, contamination-proof technique (gas chromatography of diastereomeric derivatives) that the Murchison meteorite contains at least 18 indigenous amino acids, or compounds hydrolyzable thereto. The abiotic character of this assemblage is underscored by the racemic nature of several of the amino acids, and by the fact that 12 of them are not commonly found in terrestrial proteins. These results were substantially confirmed by other authors, for 6 other C1 and C2 chondrites (Lawless et al., 1971, 1972; Oró et al. 1971a, b; Cronin and Moore, 1971, 1976; Cronin et al., 1979; Shimoyama et al., 1979). The total has grown to 37 amino acids, of which 20 are fully identified (Lawless, 1973; Kvenvolden, 1974). Interestingly, many of the protein amino acids have not been detected thus far: aromatic, hetero-aromatic, and basic amino acids, as well as those containing hydroxyl or sulfur.

The Miller-Urey reaction has been quite successful in duplicating these results. All 20 amino acids identified in meteorites, and 12 others, were produced by electric discharges on CH_4-NH_3-H_2O-H_2 mixtures, in the presence of an aqueous phase (Ring et al., 1972; Wolman et al., 1972). Even the proportions of the various amino acids resemble those in Murchison to within 1–2 orders of magnitude.

The FTT synthesis has given less impressive results, having produced only 11 definite and 8 tentative matches (Yoshino et al., 1971; Hayatsu et al., 1971). Total yields were 0.01–0.1%, much less than in the Miller-Urey synthesis (2%), though similar to the abundance in meteorites ($\leq 0.1\%$ of the organic carbon). The product distribution again was fairly similar to that in meteorites, but also included aromatic or heterocyclic amino acids such as tyrosine and histidine that cannot be made by conventional Miller-Urey syntheses. In the present context, that is a liability rather than an asset, since these amino acids have not been found in meteorites either.

These differences may reflect mainly the effort expended on the two methods, rather than their intrinsic merits. The FTT work was done before the meteorite results became available, and so many of the non-protein amino acids simply were not looked for. Also, both syntheses persumably involve the same two steps (Miller et al., 1976): formation of unstable intermediates at high T, and rapid quenching and hydrolysis of the reaction products. The standard Miller-Urey flask, with its small spark zone and large liquid phase, is an optimal configuration for this purpose, in contrast to the FTT flask, where the hot and cold zones are in reverse ratio. If the intermediates and reaction paths indeed are similar (Miller et al., 1976), then it should be possible to improve yields in the FTT synthesis merely by changing the configuration of the apparatus, to provide a larger cold zone and faster quench.

The close match of the amino acid distributions in meteorites and the Miller-Urey synthesis suggests that the meteoritic amino acids were produced by the *essential steps* of the Miller-Urey synthesis. It is not yet clear what these "essential steps" would be in a nebular setting. Synthesis of intermediates obviously does not require an electric discharge, but can also be achieved from CO, NH_3, and H_2. The required

15

extent of the gas phase reaction can presumably be achieved at lower temperatures and longer times, perhaps obviating the need for an actual quench (Anders et al., 1974). Finally, hydrolysis to amino acids may occur later, in the meteorite parent body, where an aqueous phase was present for $> 10^3$ years (Du Fresne and Anders, 1962; Bunch and Chang, 1980).

The chemical state of the amino acids in Murchison has been studied in great detail. About 80% exist in water-soluble form (Kvenvolden et al., 1971), but only in part as "free" amino acids. When extracted with D_2O, some amino acids become partially deuterated whereas others do not (Lawless and Peterson, 1975). This suggests that some amino acids either have labile H atoms, or are produced by hydrolysis of "precursors".

4.5 Porphyrins

Hodgson and Baker (1969) have detected pigments resembling porphyrins in several carbonaceous chondrites. It is not clear whether these were true (= cyclic) porphyrins or linear pyrrole polymers. Both kinds have been seen in FTT syntheses (Hayatsu et al., 1972). The cyclic pigment illustrated in Fig. 7 has a major peak at mass 580, as expected for an alkyl-substituted porphine $C_{20}D_{14}N_4 + 16\,CD_2$. Though it does not show a doubly-charged ion at mass 290, it displays other characteristics of porphyrins: strong absorption at 394 nm (in the Soret band range), formation of a red copper complex, and chromatographic and solvent extraction behavior similar to that of porphyrins (Hayatsu et al., 1972).

Porphyrin-like pigments of similar properties have been made by the Miller-Urey synthesis (Hodgson and Ponnamperuma, 1969), but they have not yet been characterized by mass spectrometry. In particular, their cyclic character has not been established.

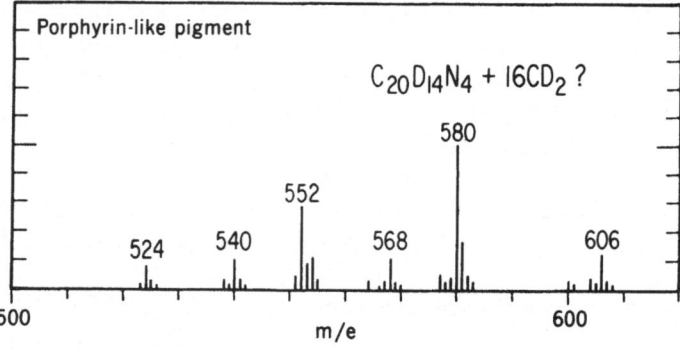

Fig. 7. Mass spectrum of porphyrin-like pigment, made from CO, D_2, and ND_3 by a Fischer-Tropsch-type synthesis (Hayatsu et al., 1972). It resembles porphyrins in optical and chemical characteristics, but lacks the expected peak at mass 290 from the doubly-charged molecular ion

4.6 Chlorine- and Sulfur-Containing Compounds

Mono- and dichlorobenzenes have been found in several carbonaceous chondrites (Studier et al., 1968, 1972), but in view of the widespread human use of such compounds, it is not all certain that they are indigenous. Benzothiophenes, first reported by Hayes and Biemann (1968) in pyrolysis experiments, have also been seen in room-temperature solvent extracts (Studier et al., 1972) and hence must be original constituents of the meteorites, not thermal degradation products. Thiophene and several of its alkyl derivatives are released from Murchison at temperatures as low as 150 °C. It is not known whether these compounds can be made in FTT syntheses, because the necessary experiments have not yet been attempted.

4.7 Organic Polymer

About 70–95% of the organic matter in carbonaceous chondrites consists of an ill-defined, insoluble macromolecular material, often referred to as "polymer" or "kerogen". A typical elemental composition for Murchison polymer (Hayatsu et al., 1980a), on a dry, ash-free basis, is C 76.5%, H 4.5%, N 2.4%, S 4.3%, and O 12.4% (by difference).

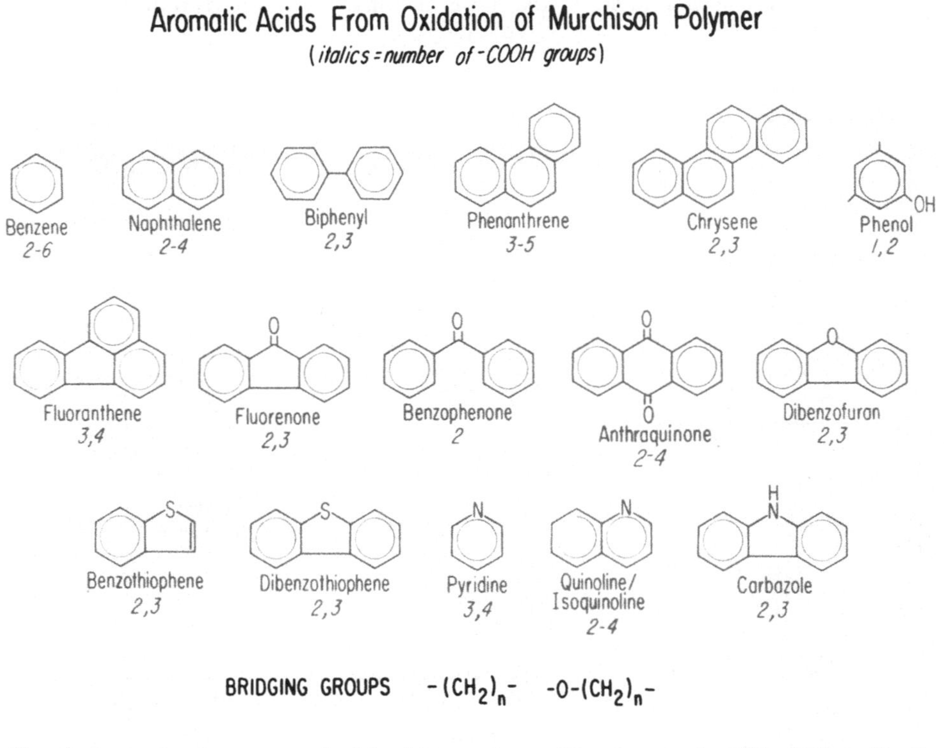

Aromatic Acids From Oxidation of Murchison Polymer
(italics = number of -COOH groups)

Benzene *2-6*	Naphthalene *2-4*	Biphenyl *2,3*	Phenanthrene *3-5*	Chrysene *2,3*	Phenol *1,2*

Fluoranthene *3,4*	Fluorenone *2,3*	Benzophenone *2*	Anthraquinone *2-4*	Dibenzofuran *2,3*

Benzothiophene *2,3*	Dibenzothiophene *2,3*	Pyridine *3,4*	Quinoline/Isoquinoline *2-4*	Carbazole *2,3*

BRIDGING GROUPS $-(CH_2)_n-$ $-O-(CH_2)_n-$

Fig. 8. Aromatic ring systems in Murchison polymer (Hayatsu et al., 1977, 1980a). Gentle oxidation converts substituents to COOH groups, but leaves ring systems intact. In addition to the ring systems shown, methyl naphthalene and methyl phenanthrene were also identified (Hayatsu et al., 1980a)

Pyrolysis of the polymer yields mainly aromatic hydrocarbons, with larger amounts of alkanes, phenols, and O-, N-, and S-containing compounds (Hayes, 1967; Bandurski and Nagy, 1976; Hayatsu et al., 1977, and references cited therein). More detailed structural information on the Murchison polymer has been obtained by oxidative degradation (Hayatsu et al., 1977, 1980a). Oxidation with a strong oxidizing agent, HNO_3, gave a series of benzene carboxylic acids. The hexa acid was the most abundant, suggesting a high degree of condensation of the aromatic nuclei in the polymer. More gentle oxidation (aqueous $Na_2Cr_2O_7$ or photochemical oxidation) gave larger fragments, representing 15 aromatic or heteroaromatic ring systems (Fig. 8).

Oxidation with alkaline CuO gave large amounts of meta-hydroxy derivatives of benzoic acid and benzene dicarboxylic acids (Hayatsu et al., 1980a). This suggests the presence of phenol ethers in the polymer structure. Interestingly, terrestrial polymers such as lignin, humic acid, and coal yield mainly *para*-rather than *meta*-hydroxy derivatives by this method.

A fairly similar material has been obtained in an FTT synthesis extended over 6 months (Hayatsu et al., 1977). Upon pyrolysis, it gave a mass spectrum resembling that of the Murchison polymer (Fig. 9). The spectrum shows mainly benzene, naphthalene, and their alkyl derivatives, as well as alkyl-indanes, fluorene, anthracene/phenanthrene, alkenes, alkanes, and alkylphenols (Hayatsu et al., 1977). This material has not been studied by the more informative, gentle oxidation methods,

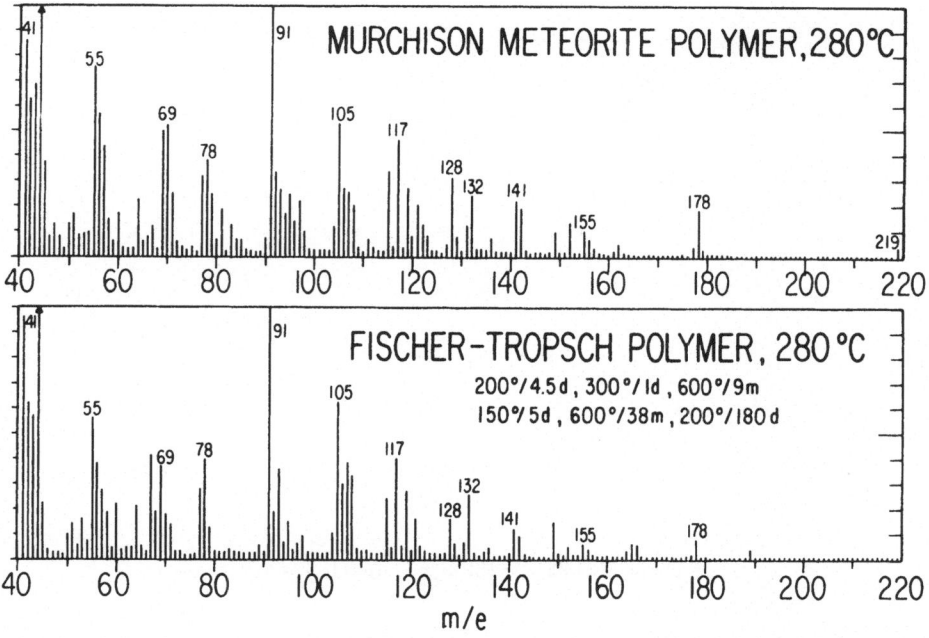

Fig. 9. Mass spectra of polymeric materials from the Murchison meteorite and a Fischer-Tropsch synthesis extended over 6 months. The principal peaks are due to aromatic hydrocarbons, their alkyl derivatives, and alkenes (Hayatsu et al., 1977)

but upon oxidation with HNO_3, it gave benzene carboxylic acids in proportions resembling those from Murchison polymer.

Polymeric materials also form in Miller-Urey reactions, by both spark discharge (Miller, 1955) and UV irradiation (Sagan and Khare, 1979). These materials have not been studied in detail, but the elemental analyses show high N contents (36% and 11%, vs. 2.4% for Murchison and 1.23% for FTT polymer). The H/C ratios also are higher (1.28 and 1.23, vs. 0.70 and 0.78), suggesting a predominantly aliphatic and/or alicyclic, rather than aromatic, structure.

4.8 Carbyne

Most of the carbon in the Allende C3V chondrite is present in elemental form, rather than as polymer or extractable organic compounds (Breger et al., 1972). It was originally called "amorphous carbon", since it is amorphous to x-rays. However, recent work shows it to be carbyne (Whittaker et al., 1980; Hayatsu et al., 1980b): a triply bonded, linear allotrope of elemental carbon. Carbyne exists in at least 10 varieties, ranging between graphite and diamond in hardness and density (Whittaker, 1978 and references therein).

At least 4 carbynes — carbon VI (?), VIII, X, and XI — have been identified in Allende on the basis on x-ray diffraction (Whittaker et al., 1980). Ion microprobe data (specifically, the predominance of even-numbered carbon fragments) suggests that at least 80% of the acid-insoluble carbon fraction is carbyne. Some of the carbynes are thermally labile, and at 250–330 °C give off various fragments, including C_{2n}^+ ($n = 1$–5) and $(C \equiv C)_n CN^+$ ($n = 1$–3). Smaller amounts of carbynes have been found in the Murchison C2 chondrite (Whittaker et al., 1980), where they are masked by the abundant organic polymer (Hayatsu et al., 1977).

Some clues to the origin of the carbynes come from their trapped noble gas components. The gases in Allende can, in principle, be derived from solar noble gases by mass fractionation, perhaps augmented by fission or some other process that produces the heavy Xe isotopes (Lewis et al., 1977). Those in Murchison, on the other hand, contain highly anomalous Ne, Kr, and Xe (Srinivasan and Anders, 1978; Alaerts et al., 1980). Both Kr and Xe show the characteristic signature of the s-process (neutron capture on a slow time scale), which apparently occurs in red giants. Presumably these carbyne grains represent stellar condensates, ejected from stars at the red giant, nova, or supernova stage (R. S. Lewis et al., 1979). Indeed, Webster (1979, 1980) has suggested that carbyne fibers, ejected from cool stars, comprise a major part of the interstellar dust.

According to a tentative phase diagram (Whittaker, 1978), carbynes are the stable forms of carbon above 2600 K, and should therefore condense in place of graphite from a carbon gas. Calculations by Clegg (1980) show, however, that the atmospheres of cool stars are too tenuous to allow carbon to condense above 2600 K; by the time the vapor becomes saturated with C, temperatures are well below 2600 K, in the stability field of graphite.

Hayatsu et al. (1980b) have shown, however, that carbyne forms metastably at much lower temperatures (520–620 K), by catalytic disproportionation of CO (to CO_2 and C) on a chromite or olivine catalyst. This reaction is much slower than the hydrogenation (FT) reaction, but the catalysts required for the latter — Fe_3O_4 and/or clays — do not form in a solar gas until T has fallen to ≤ 400 K. Above that temperature, the dominant mineral is olivine, and the dominant reaction is the slow disproportionation of CO to carbynes.

These relations are illustrated in the "phase diagram" in Fig. 10. C3 chondrites, having formed above the 400 K threshold for formation of hydrogenation catalysts (Anders et al., 1976), contain carbynes but little or no polymer or organic compounds.

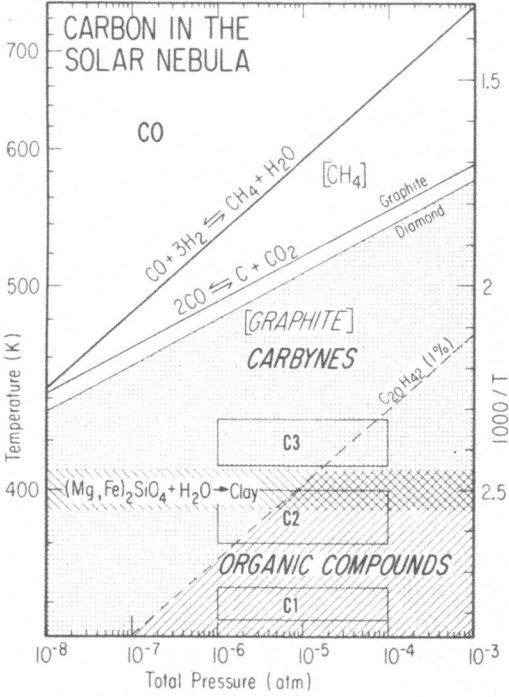

Fig. 10. Chemical state of carbon in the solar nebula (Hayatsu et al., 1980b). Solid lines give the temperatures at which 50% of the CO should have reacted according to the equilibrium shown (the 0.1% lines would be typically 100–200° higher). In each field, principal stable products are shown in roman type, and metastable products, in italics; those that do not form for kinetic reasons are enclosed in brackets.

For example, at 10^{-5} atm, 50% of the CO should have transformed to CH_4 at 590 K, but this reaction is very slow in the absence of catalysts and so CO may instead disproportionate to CO_2 and elemental carbon. This reaction should be 50% complete at 520 K. It yields carbynes rather than graphite, if a chromite catalyst is present. At 400 K, clays form from anhydrous silicates, catalyzing hydrogenation of CO to complex organic compounds. The dashed line shows the temperatures at which 1% of the CO transforms to a typical alkane, $C_{20}H_{42}$, under metastable conditions (the lines for most other alkanes, from C_3H_8 upward, are very similar).

Chemical state of carbon in carbonaceous chondrites agrees with that predicted from their formation conditions (indicated by boxes), as inferred from isotopic fractionation of O and C, or abundances of volatile metals (Table 1 and Fig. 11; Onuma et al., 1972, 1974; Anders et al., 1976). C1 and C2 chondrites, having formed between 360 and 400 K, contain mainly organic compounds with only traces of carbynes (Whittaker et al., 1980). C3 chondrites contain mainly elemental carbon, which, at least in the case of Allende, is present as carbynes rather than graphite

C1 and C2 chondrites, having formed below the 400 K threshold, contain large amounts of organics.

No attempt has yet been reported to produce carbynes by the Miller-Urey reaction. This should not be held against it, since carbynes have only very recently been discovered in meteorites. At least acetylene and some of its simpler derivatives have been made in the Miller-Urey synthesis (Friedman et al., 1971).

5 Isotopic Data

5.1 Carbon

Meteorites show a very large difference in C^{12}/C^{13} ratio between carbonate and organic carbon: 60 to $80\%_{00}$ (Boato, 1953; Briggs, 1963; Clayton, 1963; Smith and Kaplan, 1970). This trend remained unexplained for a number of years, because coexisting carbonate and organic matter on Earth shows a much smaller difference, typically $25–30\%_{00}$. It is a primary feature, unaffected by the later thermal history of the meteorite. Terrestrial calcium carbonate is not known to equilibrate with coexisting organic matter in sediments (Smith and Kaplan, 1970).

Although fractionations of $60–80\%_{00}$ are theoretically possible under equilibrium conditions at very low temperatures ($\leqq 0$ °C), they are not observed on Earth. Urey (1967) therefore proposed that the two types of carbon came from two unrelated reservoirs, whereas Arrhenius and Alfvén (1971) suggested fractionation during carbonate growth from the gas phase, involving multiple desorption or metastable molecules.

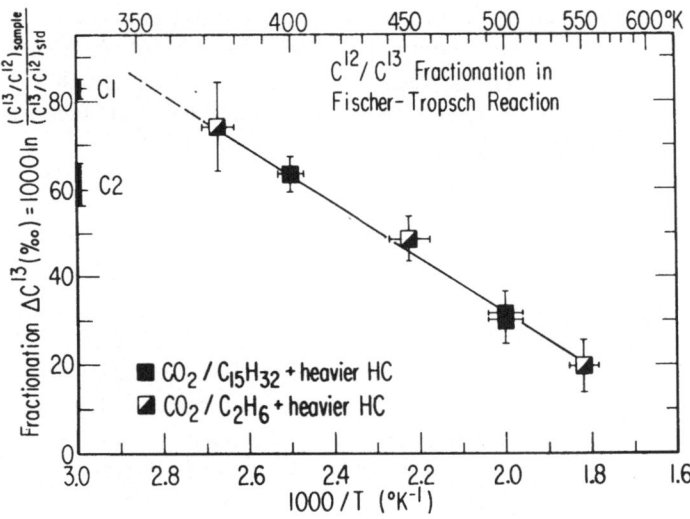

Fig. 11. The Fischer-Tropsch reaction shows a kinetic isotope fractionation between organic and carbonate carbon of the same sign and magnitude as that in meteorites (black bars on ordinate). Observed fractionations in C1 and C2 chondrites correspond to temperatures of about 360 K and 400 K (Lancet, 1972)

It turns out, however, that the Fischer-Tropsch reaction gives an isotopic fractionation of just the right sign and magnitude, owing to a kinetic isotopic effect (Lancet and Anders, 1970; Lancet, 1972). The temperature dependence of the fractionation between 375 and 500 K suggests that the observed fractionations in C1 and C2 chondrites correspond to about 360 to 400 K (Fig. 11). These values agree rather well with the formation temperatures of carbonates and silicates, based on O^{18}/O^{16} ratios, 360 K for C1's and 380 K for C2's (Onuma et al., 1972, 1974). The Miller-Urey reaction gives a fractionation of only $-0.4 \pm 0.2\%_0$ (Lancet, 1972).

This interpretation has been questioned by Chang and Mack (1978), who found that water-soluble organics, especially amino acid extracts from the Murchison meteorite, were almost as heavy as carbonate C [δC^{13}(PDB) = +23 to +44$\%_0$ vs. +44.4$\%_0$], in contrast to the much lighter insoluble polymer (-15.3 to $-16.8\%_0$) or benzene-methanol extracts ($+5.0\%_0$). They suggested that these various forms of carbon represent several stages of carbon condensation in the solar nebula, "in different environments separated in space and possibly in time".

This explanation gains support from isotopic data for N and O (Sec. 5.2 and 5.3), which also point to an isotopically heterogeneous nebula. However, the amino acid data actually are not inconsistent with the FTT reaction. Hydrogenation of CO can yield either CO_2 or H_2O:

$$2\,CO + 2\,H_2 \rightarrow CH_4 + CO_2 \qquad \Delta G^\circ = -59{,}100 + 30.1\,T\ \text{(cal/mole)} \qquad (1)$$
$$CO_2 + 4\,H_2 \rightarrow CH_4 + 2\,H_2O \qquad \Delta G^\circ = -39{,}420 + 41.2\,T\ \text{(cal/mole)} \qquad (2)$$

$$CO + 3\,H_2 \rightarrow CH_4 + H_2O \qquad \Delta G^\circ = -49{,}260 + 51.2\,T\ \text{(cal/mole)} \qquad (3)$$

For kinetic reasons, reaction (3) normally dominates. However, reaction (2) still is exoergic, and so any CO_2 formed by (1) can subsequently hydrogenate according to (2), forming a second crop of organic compounds (Anderson, 1956). And since the CO is isotopically heavy (Fig. 11), the resulting organic compounds will also be heavy. Thus "primary" organic compounds from CO will be light, whereas "secondary" compounds, from CO_2, will be heavy.

5.2 Nitrogen

The nitrogen data, while more ambiguous, do not fit this picture. Organic nitrogen in C1 and C2 chondrites has δN^{15} of $+30$–$50\%_0$, relative to atmospheric N_2 (Kung and Clayton, 1978). No second, co-existing N-phase has been analyzed in these meteorites, but inorganic N in enstatite chondrites — located in high-temperature minerals and hence probably approximating the solar nebula — has δN^{15} of -30–$40\%_0$, for a total difference of 80–90$\%_0$ (Kung and Clayton, 1978).

Chemical fractionation apparently cannot account for this difference. Kung et al. (1979) have found that N-isotope fractionations in both FTT and Miller-Urey reactions are too small: $+3\%_0$ and $+10$–$12\%_0$. The high δN^{15} of C1 and C2 chondrites could, in principle, be explained by mass fractionation in a Rayleigh process, involving loss of 99% of the N_2. But this process would have to be driven to ridiculous extremes to account for the δN^{15} of $+170\%_0$ of the C2R chondrite Renazzo (Kung and Clayton, 1978).

The most plausible explanation remaining is isotopic heterogeneity of the solar nebula: C- and E-chondrites may have originated in isotopically distinct regions, differing in δN^{15} by 80–90$\%_0$ (Kung and Clayton, 1978). A similar explanation has already been invoked for oxygen (R. N. Clayton, 1978), where the existence of 3 rather than 2 stable isotopes makes the interpretation less ambiguous. *Chemical* heterogeneity has long been implied by the highly reduced mineralogy of E-chondrites, requiring a high C/O ratio (Larimer, 1968, 1975).

Of course, if the isotopic variations in N and O are due in part to heterogeneities of the nebula, then one must admit this possibility for C as well. But the isotopically distinct carbon phases — carbonate, polymer, and amino acids — all coexist in the same meteorite, which makes origins in different regions rather less probable. Still, this possibility must be kept in mind.

5.3 Hydrogen

Hydrogen in C1 chondrites occurs in three major forms: (1) loosely bound water of hydration of $MgSO_4$ and other salts; (2) structural water in clay minerals, and (3) organic hydrogen. Boato (1953) found in a stepped heating experiment on the C1 chondrite Ivuna that the water released up to 180 °C was of essentially terrestrial composition, and apparently represented exchangeable water of hydration, whereas the water released between 180° and 900 °C became progressively heavier (δD up to 42%, relative to standard mean ocean water), and apparently came from the other two, indigenous, sources. A similar experiment on the C2 chondrite Cold Bokkeveld, on the other hand, gave water of more normal and less variable composition.

The first attempt to determine the isotopic composition of the organic fraction was made by Briggs (1963). He found that benzene-methanol extracts of four C1, C2, and C3V chondrites had variable isotopic compositions, in each case closely resembling the δD and δC^{13} values measured by Boato (1953) on bulk samples of the same meteorites. This implied that the organic H was isotopically similar to the more abundant inorganic H in the silicates, which dominated the bulk samples of Boato.

Very different results have been found in recent studies, by stepped heating (Robert et al., 1979, 1980), oxidation in an oxygen plasma (Kolodny et al., 1980), or prior enrichment of organic matter by solvent extraction or removal of inorganic minerals with HCl—HF (Robert et al., 1980; Robert and Epstein, 1980). The organic matter shows very large deuterium enrichments, up to $\delta D = 150\%$ for Orgueil and 310% for Renazzo (Robert and Epstein, 1980). C and N also vary, but only within the previously observed range, and do not correlate with D. The hydrogen in clay minerals, on the other hand, is very much lighter: $\delta D = +3$ to -16% for Orgueil, Murray, and Murchison; $+68\%$ for the perpetual non-conformist, Renazzo.

Various possible mechanisms have been discussed by the above authors, and — most comprehensively — by Geiss and Reeves (1981). Equilibrium chemical fractionations in the solar nebula could possibly account for the silicates, which require temperatures around 230–300 K (Kolodny et al., 1980; Robert and Epstein,

1980), but hardly for the organic compounds: at the required temperatures of 130–180 K, reaction rates are negligibly slow. Kinetic isotope effects are altogether unsuitable, because they have the wrong sign. Spallation reactions, in turn, are ruled out by overproduction of light xenon isotopes.

The principal possibility remaining is that the deuterium enrichment was caused by ion-molecule reactions, analogous to those postulated for interstellar molecules. It is well known that certain interstellar molecules such as HCO^+, HCN, or HNC are enriched in D by factors of 10^2–10^5 (Penzias, 1979; Snell and Wootten, 1979). These enrichments are attributed to ion-molecule reactions at low temperatures, which cause D to concentrate in ionic, reactive species such as H_2D^+ or CH_2D^+ (Watson, 1976; Huntress, 1977; Snell and Wootten, 1979). The D-enrichment in meteoritic organic matter is up to four orders of magnitude smaller than in interstellar molecules, and so the meteoritic organic matter can't simply be mummified interstellar molecules. It probably consists mainly of local material that was slightly contaminated with D-rich, interstellar material. A telling clue (Geiss and Reeves, 1981) is the observation that these very same meteorites contain one or more exotic noble gas components in carbonaceous host phases, whose isotopic compositions clearly point to a presolar origin (Eberhardt et al., 1979; R. S. Lewis et al., 1979).

On the other hand, one cannot rule out the possibility that D-rich, ionic species formed at some stage in the solar nebula, and reacted with previously-produced, polymeric material. This is essentially a Miller-Urey reaction with a built-in, isotopic tracer. Perhaps these two alternatives can be distinguished by isotopic analysis of carefully separated fractions of the organic material.

6 Feasibility of the FTT Synthesis in the Solar Nebula

It appears that FTT reactions can account reasonably well for most features of organic matter in meteorites. The only alternative process, the Miller-Urey synthesis, fails to account for the aliphatic and aromatic hydrocarbons, nitrogen heterocyclics, many oxygen compounds, the polymer, and carbon isotope fractionations, though it remains an alternative and perhaps superior source of amino acids and may, in an extended sense, be responsible for the deuterium enrichments.

However, it is not immediately obvious that the FTT model experiments are relevant to the solar nebula. First and foremost, it must be shown that CO was still present at the time the nebula had cooled to 360 K.[3] How did CO traverse the no-man's land between 600 K where it becomes unstable with respect to CH_4, and ~400 K, where formation of heavy hydrocarbons first becomes thermodynamically feasible (Fig. 1, 10)? Second, was the reaction rate fast enough at the extremely low pressures in the nebula?

---

[3] We are making the conventional assumption that the nebula cooled isobarically from high temperatures. Actually, the same final state could be reached by isothermal compression or by some intermediate path (Arrhenius and Raub, 1978). In that case, survival of CO would be virtually assured. CO is the stable from of C at low P and T (Fig. 10), and although it would have to traverse the "graphite-carbyne" field in Fig. 10 before reaching the "organic compound" field, most of it would survice, since reaction rates at low pressures are very slow (Sect. 6.2).

6.1 Survival of CO

An answer to the first question was suggested by Lancet and Anders (1970). The principal meteoritic phases stable above $\sim 350\text{--}400$ K (olivine, pyroxene, Fe, FeS) are not effective catalysts for the Fischer-Tropsch reaction, whereas the phases forming below this temperature (hydrated silicates, magnetite) are. [Though metallic iron is often regarded as a catalyst for this synthesis, the catalytically active phase actually is a thin coating of Fe_3O_4 formed on the surface of the metal (Anderson, 1956)]. Thus CO may have survived metastably until catalysts became available by reactions such as:

$$12 \, (Mg, Fe)_2 SiO_4 + 14 \, H_2O \rightarrow 2 \, Fe_3O_4 + 2 \, H_2 +$$
$$+ \, 3 \, (Mg, Fe)_6 (OH)_8 Si_4 O_{10}$$
$$4 \, (Mg, Fe)_2 SiO_4 + 4 \, H_2O + 2 \, CO_2 \rightarrow 2 \, (Mg, Fe)CO_3 +$$
$$+ \, (Mg, Fe)_6 (OH)_8 Si_4 O_{10} \, .$$

This would also explain why the hydrated silicates, carbonates, and organic compounds in C1's all have the same formation temperature (Table 1). (The deuterium enrichment (Sect. 5.3), if interpreted as an equilibrium fractionation, suggests lower temperatures, but it may actually be due to reactions with D-rich ions).

More recently, Lewis and Prinn (1980) have systematically examined the reduction kinetics of CO and N_2 in the solar nebula. Taking into account gas-phase reactions as well as surface-catalyzed reactions (for the rather inefficient catalysts present above 400 K), they conclude that reaction rates were so slow relative to the rates of radial mixing or nebular evolution that no more than 1% of the N_2 and CO would have been reduced to NH_3 and CH_4 over the lifetime of the nebula. Methane, the starting material of the Miller-Urey reaction, apparently was only a minor constituent of the solar nebula.

6.2 Rate of Reaction in the Nebula

The second question cannot be answered unequivocally because the kinetics of the Fischer-Tropsch reaction is not well enough understood to permit reliable extrapolations to very low pressures. Still, a tentative analysis based on Langmuir's adsorption isotherm suggests that the rate may be adequate even at 10^{-5} atm (Lancet, 1972).

Since H_2 is much more strongly adsorbed than is CO and covers nearly the entire surface at pressures greater than 10^{-10} atm (Hayward and Trapnell, 1964), the Langmuir expression for the rate R simplifies to:

$$R = k(b_C/b_H) \, (P_C/P_H^{1/2})$$

where P = pressure, k = rate constant, b = adsorption coefficients and the subscripts C, H stand for CO and H_2. (The exponent of 1/2 reflects the fact that H_2

25

dissociates to H atoms upon adsorption). With symbols N = nebula and L = laboratory, the half-time t in the nebula is:

$$t_N = t_L P_{CL} P_{HN}^{1/2} / P_{CN} P_{HL}^{1/2} \qquad (4)$$

A conservative estimate of t_L may be obtained from the experiments of Lancet (1972), who studied the rate of the Fischer-Tropsch reaction on a cobalt catalyst between 500 and 375 K (Fig. 12). The data give an activation energy of 27 ± 1 kcal/mole, similar to earlier determinations (Anderson, 1956). The laboratory half-time t_L extrapolated to 360 K is 8600 hr, or ~ 1 yr. With $P_{HL} = P_{CL} = 0.5$ atm, $P_{HN} = 10^{-5}$ atm, and $P_{CN} = 6.5 \times 10^{-9}$ atm, we obtain $t_N = 3.4 \times 10^5$ yr.

However, this value almost certainly is a gross overestimate. Lancet's experiments were done in a static system at 1 atm, where the mean distance between gas molecules and catalyst was ~ 10 cm or $\sim 10^6$ mean free paths. In the nebula, this distance would be only ~ 0.03 cm or ~ 0.01 mean free paths.[4] A better estimate

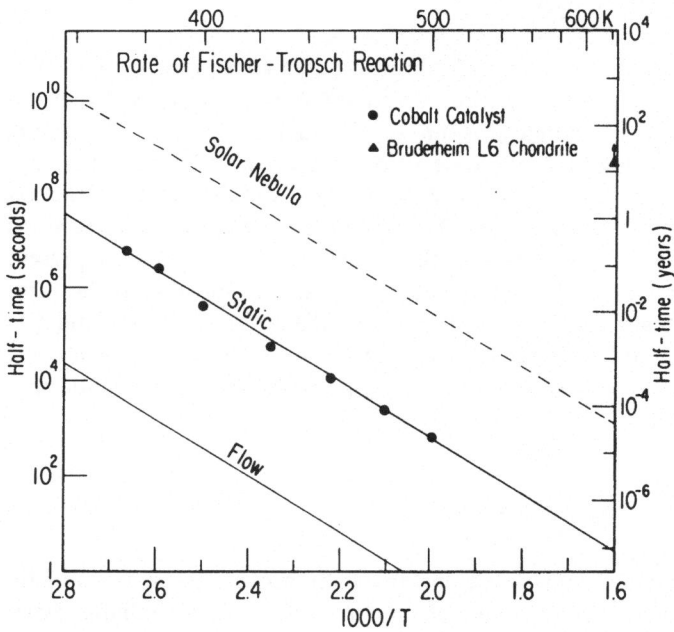

Fig. 12. Fischer-Tropsch reaction at 1 atm is first-order in CO, with an activation energy of 27 kcal/mole (Lancet, 1972). Rate in a flow system is $\sim 10^3$ times faster than in the static system used here. Dashed line shows extrapolation to solar nebula, assuming that the rate is proportional to $(P_{CO})(P_{H_2})^{-1/2}$. Reaction proceeds at an undetectable rate when the Bruderheim L6 chondrite is used as a catalyst. Apparently the high-temperature minerals in this meteorite (olivine, orthopyroxene, troilite, and nickel-iron) do not catalyze the hydrogenation of CO. Thus CO can survive in the solar nebula down to 400 K, when catalytically active minerals first from (Fig. 1 and 10)

---

[4] The dust density in a solar gas at 360 K and 10^{-5} atm is 5.65×10^{-12} g/cm^3, assuming Cl chondrite composition. With a grain radius of 10^{-5} cm and density 2.4 g/cm^3 (typical values for Cl chondrites), the number density is 565 grains/cm^3, corresponding to a mean distance of 0.12 cm between grains, or 0.03 cm between gas molecules and grains.

may therefore be obtained from industrial flow-type syntheses, where contact between gas and catalyst is more intimate than in a static system. Rates at 440 to 470 K (Anderson, 1956) are typically $(1-2) \times 10^3$ times faster (bottom curve in Fig. 12) than in Lancet's static experiments (middle curve). Accordingly, the half-time in the nebula (top curve) may be 500 yr or less, somewhat shorter than the expected lifetime of the nebula of $\sim 10^5$ yr. It would be shorter still at higher nebular pressures $(10^{-5}-10^{-3}$ atm), which are currently favored by many authors.

We can reexamine the question of CO survival in the light of these rate data. From the relation between surface density and cooling time (Cameron, 1962), it seems that the pertinent region of the nebula took 10 years or 3×10^8 sec to cool from 600 to 400 K. The requirement that one-half the initial CO survive this stage is equivalent to the condition that the half-time at 600 K, in the *absence* of good catalysts, be greater than 3×10^6 sec. This is nearly three orders of magnitude longer than the halflife of 5×10^3 sec extrapolated from the top curve in Fig. 12, which applies in the *presence* of good catalysts. The available experimental data do indeed suggest a differences of at least this order. No trace of a reaction was observed withn CO and D_2 were heated at 620 K in the presence of the L6 chondrite Bruderheim (Studier et al., 1968), which represents the mineral assemblage expected in the nebula between 600 and 400 K. The lower limit to the half-life inferred from this experiment is 4×10^8 sec (triangle in Fig. 12), i.e. 8 orders of magnitude above the line for a Co catalyst. It appears that the survival of CO between 600 and 400 K is not a problem.

We must also check whether enough NH_3 was present in the nebula for synthesis of nitrogen compounds. At equilibrium, only about 1% of the total N will be present as NH_3 at 360 K and 10^{-5} atm, and even less at lower pressures (Fig. 1). Kinetic considerations give similar values, around 1% (Lewis and Prinn, 1980; Norris, 1980). However, Cl chondrites contain only 1.5% of their cosmic complement of N, and so even $\sim 1\%$ in the form of NH_3 may have sufficed.

6.3 Catalysts

The availability of a catalyst is another important question. Industrial Fischer-Tropsch syntheses generally use metallic catalysts, and since these are easily poisoned by sulfur, Oró et al. (1968) have questioned the catalytic effectiveness of meteoritic dust in the solar nebula, in view of the high sulfur content of meteorites.[5]

However, this objection does not seem serious (Anders et al., 1973). In a poisoned industrial catalyst, the entire surface is coated with sulfide, rendering it ineffective. But in carbonaceous chondrites, the catalytically active magnetite and phyllosilicate grains are discrete and spatially separated from the sulfides. Grains suspended in the nebula cannot intercommunicate with each other, and hence a sulfide grain cannot inhibit catalytic reactions at the surface of an Fe_3O_4 grain 0.1 cm distant.

---

[5] Miller et al. (1976) have repeated the argument about sulfur poisoning, and have raised several other objections to the FTT synthesis, without mentioning that they had been extensively discussed and refuted several years earlier (Anders et al., 1973, and earlier papers of the Chicago group).

Moreover, phyllosilicate is immune to sulfur poisoning. Indeed, Studier et al. (1968) showed that the Cold Bokkeveld carbonaceous chondrite, despite its sulfur content of 3%, was able to catalyze an FTT reaction.

There are some indications that the organic compounds were synthesized on the surfaces of meteoritic mineral grains. Vdovykin (1967) and Alpern and Benkheiri (1973) have shown that much of the organic matter is present as rounded, fluorescent particles 1–3 μm in diameter, containing cores of magnetite or silicate.

7 Interstellar Molecules

Lederberg and Cowie (1958) suggested long ago that complex organic molecules might form in space, by gas phase reactions or surface catalysis on dust grains. Since then, more than 50 interstellar molecules or radicals have been discovered (Gammon, 1978), and until recently, it appeared that both mechanisms played a role: gas-phase, especially ion-molecule reactions for at least the smaller molecules, and grain catalysis for H_2 and perhaps the larger molecules (Watson, 1976; Allen and Robinson, 1977; Gammon, 1978). Since then, there has been a decided shift in favor of ion-molecule reactions, not because grain catalysis looks worse but because ion reactions look better. The large D enrichments (Sec. 5.3) clearly point to ion reactions, and some new reaction schemes endeavor to build even the largest molecules by ion reactions (Smith and Adams, 1978; Huntress and Mitchell, 1979; Schiff and Bohme, 1979). We shall review the problem with an admitted bias toward grain catalysis.

7.1 Synthesis in Solar Nebulae?

Herbig (1970) suggested that solar nebulae might be manufacturing sites for interstellar molecules. Main sequence stars show a marked discontinuity of rotation rates at 1.5 $M_\odot$, suggestive of angular momentum transfer to extrastellar material. Thus solar nebulae may be a common byproduct of star formation. Such nebulae, embedded in interstellar clouds, provide a high-density environment ($\sim 10^{14}$ molecules/cm³) in which matter can be transformed to grains and molecules that are then returned to the interstellar cloud when the nebula is dissipated.

In our own solar system, nearly all volatiles complementary to the inner planets (3×10^{-3} $M_\odot$) were so lost. Earth and Venus contain only about 10^{-4} their complement of C, and even lesser amounts of H_2O, N, and noble gases. Since the retained C appears to show the imprint of the Fischer-Tropsch reaction, it seems likely that the lost C, too, had been involved in this process.

There exists some circumstantial evidence linking interstellar molecules to protostars. The more complex interstellar molecules tend to occur only in regions of very high density ($\geqq 10^4$ H_2/cm³), e.g. the infrared nebula in Orion. Star formation proceeds at a rapid rate in such clouds, and thus solar nebulae may have formed and dissipated. The lifetime of our solar nebula seems to have been rather short: 10^5 to 10^6 yr, or 1–2 orders of magnitude less than the age of a typical cloud, e.g. the Orion Nebula, or lifetimes of molecules against UV photolysis (Gammon, 1978).

If these values are typical, even a young cloud should contain appreciable amounts of carbon cycled through solar nebulae. Abundances of interstellar molecules relative to CO are at least 2 orders of magnitude lower than yields in FTT syntheses (Gammon, 1978). It appears that only a moderate degree of star formation and CO processing would suffice to account for the interstellar molecules.

An apparent limitation of the FTT reaction is its high activation energy of 27 kcal/mole, which requires temperatures above 300 K for reasonable rates. (For example, at $P = 10^{-10}$ atm, the temperature for 1% conversion to heavy alkanes is 294 K, and the corresponding time, 6.8×10^6 yr). However, the rate measurements extend only to 375 K, and do not preclude a second pathway of low or zero activation energy that becomes dominant at lower temperatures. Adsorption of the reactants itself involves no activation energy (Brecher and Arrhenius, 1971), and the activation energy associated with the reaction or desorption steps can be supplied or reduced in varous ways: more active reactants, e.g. H instead of H_2, heating of the grain by exothermic reactions (Allen and Robinson, 1977), or high surface coverage of the grain, which decreases the desorption enthalpy (Brecher and Arrhenius, 1971). Thus there is at least some chance that the FTT reaction may work at the low temperatures and pressures of interstellar clouds.

7.2 Observations

Of the 31 known interstellar molecules with ≥ 3 atoms (Table 2), at least 23 have been found in meteorites or FTT syntheses. They are italicized in Table 2.

One feature suggestive of the FTT reaction is the variable abundance of the two structural isomers, CH_3OCH_3 and C_2H_5OH, which dominate in the Orion Nebula and Sagittarius B2, respectively (Zuckerman et al., 1975). A similar variation was

Table 2. Interstellar Molecules with 3 or More Atoms (Gammon, 1978)

C and O	N	S
CH_4 (?)	HCN	H_2S
$\{$ $HC \equiv CH$	HNC	SO_2
$\{$ $CH_3C \equiv CH$	$HNCO$	OCS
		H_2CS
H_2O	$\{$ CH_3CN	
$\{$ $HCHO$	$\{$ C_2H_5CN	
$\{$ CH_3CHO	$CH_2 = CHCN$	
$H_2C = CO$	NH_3	
$HCOOH$	CH_2NH	
$HCOOCH_3$	CH_3NH_2	
$\{$ CH_3OH	NH_2CN	
$\{$ C_2H_5OH	NH_2CHO	
CH_3OCH_3	$\{$ $HC \equiv CCN$	
	$\{$ $H(C \equiv C)_2CN$	
	$\{$ $H(C \equiv C)_3CN$	
	$\{$ $H(C \equiv C)_4CN$	

seen in the FTT synthesis, where the ether/alcohol ratio varied from 0.05 to 3 for nickel-iron and clay catalysts, respectively (Anders et al., 1974). But ion reactions probably can also give a similar variation (Huntress and Mitchell, 1979).

A more general hallmark of the FTT reaction is the presence of homologous series: aldehydes, alcohols, nitriles, and cyanoacetylenes (Table 2). The FTT reaction builds carbon chains by successive addition of C_1 or C_2 units, and so the lightest member of a series is always accompanied by its heavier homologues. A good test case is provided by the cyanoacetylenes: the most numerous family, with the largest member (HC_9N, with 11 atoms).

The discovery of a large deuterium enrichment in HC_3N in the Taurus Molecular Cloud 1 ($DC_3N/HC_3N = 0.02-0.08$; Langer at al., 1980) clearly shows that ion-molecule reactions were involved at some stage. But this does not prove that the entire molecule is built step-by-step by ion-molecule reactions. Several ingenious schemes involving such buildup have been proposed (Huntress, 1977; Churchwell et al., 1978; Mitchell and Huntress, 1979; Schiff and Bohme, 1979; Walmsley et al., 1980). But all are based on a judicious choice of reactants (such as C_2H_2) and reactions that favor formation of cyanoacetylenes. Only a limited number of competing or destructive reactions has been considered, and in view of the potentially vast number of such reactions, it is not at all clear that the basic problem has been solved: how to build and preserve highly unsaturated molecules such as HC_9N in a cosmic gas, where the H/C ratio is $\sim 10^3$. Also, one of the two main reactions invoked for buildup of heavier cyanoacetylenes:

$$HC_3N + C_2H_2^+ \rightarrow H_2C_5N^+ + H$$

does not occur at a measurable rate in the laboratory; instead, $H_2C_3N^+$ forms by proton transfer (Freeman et al., 1978, 1979).

It also seems unlikely that the proposed reaction schemes for cyanoacetylenes can be stretched to produce the structurally similar but much larger, presolar carbynes in meteorites (Whittaker et al., 1980; R. S. Lewis et al., 1979). Being presolar, they, too, need to be accounted for from now on.

To its embarrassment, the FTT reaction has failed to yield the heavier cyanoacetylenes in laboratory syntheses (Anders et al., 1974). However, this probably reflects not an intrinsic shortcoming but merely destruction of the very reactive, heavier members by secondary reactions under the extreme conditions of these experiments (P = 2 atm, T = 570 K, t = 168 hr). Given the universal formation of homologous series in the FT reaction, and its proven ability to make higher carbynes (Hayatsu et al., 1980b), there is little reason to doubt that it also makes the higher cyanoacetylenes. (But so do gas phase reactions, given favorable reactants such as C_2H_2 and HCN. Winnewisser et al., (1978) produced HC_3N and small amounts of HC_5N by RF discharge in such a gas mixture at 0.1–1 torr.)

Let us compare abundance patterns in detail (Fig. 13), using the most complete sets of data available: alcohols for the FTT reaction (Anders et al., 1974) and cyanoacetylenes in TMC-1 for interstellar molecules (Morris et al., 1976; Kroto et al., 1978; Broten et al., 1978; Langer et al., 1980).

In the FTT reaction, the abundance ratio of successive homologues remains constant: $C_{n+1}/C_n = a$, and so a logarithmic plot of abundance against carbon

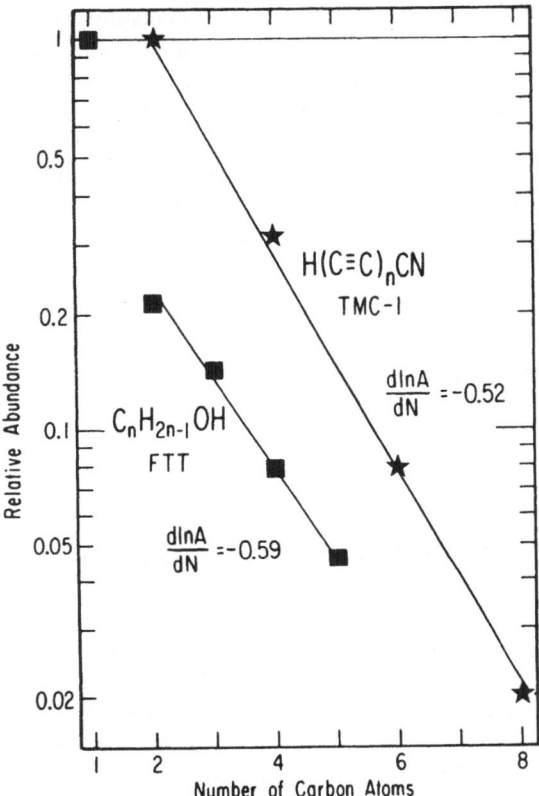

Fig. 13. The Fischer-Tropsch reaction produces homologous series in logarithmic distribution, as illustrated here by alcohols (Anders et al., 1974). From C_2 upward, successive homologues form in fixed, molecular ratio: $C_{n+1}/C_n = a$, where a is the probability of chain growth, typically 0.6–0.9. Interstellar cyanoacetylenes in the Taurus Molecular Cloud 1 likewise show a logarithmic distribution, of very similar slope (0.52 vs 0.59 for the FTT alcohols). Presumably they, too, formed by surface catalysis, rather than by gas-phase reactions

number is a straight line, of slope a (Fig. 13). The slope a simply is the probability that the carbon chain will grow by one unit. In industrial syntheses, a is typically 0.05–0.9 for $n = 1$, and 0.6–0.9 for $n \geq 2$. The value for the FTT alcohols in Fig. 13 is 0.59.

The slope for the interstellar cyanoacetylenes is quite similar: 0.52. Here the chain grows by C_2 units, and so the probability of forming the next higher homologue is $(0.52)^2$, or 0.27. In other words, conversion to the next homologue is one-third as likely as all other fates combined: hydrogenation, branching, dissociation, ionization, chain termination, etc. It seems unlikely that such remarkably high growth probabilities — moreover, constant for all 4 homologues — can be achieved in a gas phase environment. It remains to be shown, by a truly comprehensive review of *all* relevant reactions, that a net growth probability of 0.27 can be achieved in a fundamentally destructive gas phase.

It seems that the FTT reaction is better able to account for the data, since it is known to yield a distribution of the observed shape and slope. Nonetheless, the high abundance of DC_3N in TMC-1 suggests that reactions with D-rich species did occur at some stage — either on grain surfaces during the initial synthesis, or later on, by exchange reactions in the gas phase.

Thus a combination of the two processes may be needed for cyanoacetylenes: grain catalysis for formation of carbyne chains (with or without H and N) and ion reactions for hydrogenation or deuterium exchange. It is interesting that a meteoritic carbyne sample heated at only 530 K releases molecules in the right mass range, yielding C_2^+–C_{10}^+ and CN^+–C_7N^+ ions in the mass spectrometer (Hayatsu et al., 1980b). Very similar results were obtained on synthetic carbynes, made by catalytic disproportionation of CO. The low release temperature shows that these molecules or radicals are present as such in the sample, and are held in place only by weak, adsorptive (?) forces. Consequently, they may be released from the grains by the mechanisms discussed by Allen and Robinson (1977).

8 Origin of Volatiles on Inner Planets

8.1 Accretion of Planets

Some of the most volatile substances occur in the Earth's crust in nearly the same proportions as in carbonaceous chondrites. This has led to the suggestion that the inner planets obtained their volatiles from carbonaceous-chondrite-like dust or larger bodies that had formed at later times or in cooler regions of the nebula (Anders, 1968; Turekian and Clark, 1969; Anders and Owen, 1977). Such material might have made a significant contribution to the Earth's initial endowment of organic matter. Bodies of meteoritic size (1–100 cm) would survive atmospheric entry, and deliver their organic compounds intact. Larger bodies would vaporize on impact, causing any organic compounds to revert to CO and H_2. But on expansion and cooling of the gas ball, catalytic reactions might commence on the surfaces of dust grains. Apart from prebiotic organic matter, some portion of the Earth's petroleum may have originated by this route, as argued by Robinson (1966), Porfir'ev (1971), and others.

8.2 FTT Reactions on the Earth

There are indications that FTT reactions also take place in petroleum deposits. Friedel and Sharkey (1963) have shown that the alkane isomer distribution up to C_9 in crude oils can be quantitatively represented by a formula first devised for the Fischer-Tropsch reaction, in which the only parameter is the ratio of the probabilities of chain branching, b, and chain lengthening, a; $f \equiv b/a$. The resemblance of the ostensibly biogenic Nonesuch shale hydrocarbons to Fischer-Tropsch hydrocarbons (Fig. 2) suggests that this similarity holds also at higher carbon numbers. The most straightforward explanation of the quantitative applicability of the Fischer-Tropsch formula is that Fischer-Tropsch type, surface-catalyzed reactions are involved in the formation of petroleum, at least under more extreme conditions of diagenesis. Several authors (Pikovskiy et al., 1965; Galwey, 1969; Hayatsu et al., 1971; Anders et al., 1974) have shown that clay minerals are good catalysts for such reactions.

9 Acknowledgement

This work was supported in part by the Office of Basic Energy Sciences, Division of Chemical Sciences, D.O.E. (Ryoichi Hayatsu), and NASA Grant NGL-14-001-010 (Edward Anders).

10 References

Alaerts, L., Lewis, R. S., Matsuda, J. and Anders, E.: Geochim. Cosmochim. Acta 44, 189–209 (1980).
Allen, M. and Robinson, G. W.: Astrophys. J. 212, 396–415 (1977).
Allen, W. V. and Ponnamperuma, C.: Currents in Mod. Biol. 1, 24–28 (1967).
Alpern, B. and Benkheiri, Y.: Earth Planet. Sci. Lett. 19, 422–428 (1973).
Anders, E.: Acc. Chem. Res. 1, 289–298 (1968).
Anders, E.: Ann. Rev. Astron. Astrophys. 9, 1–34 (1971).
Anders, E., Hayatsu, R., and Studier, M. H.: Science 182, 781–790 (1973)
Anders, E., Hayatsu, R., and Studier, M. H.: Astrophys. J. 192, L101–L105 (1974).
Anders, E., Higuchi, H., Ganapathy, R. and Morgan, J. W.: Geochim. Cosmochim. Acta 40, 1131–1139 (1976).
Anders, E. and Owen, T.: Science 198, 453–465 (1977).
Anderson, R. B.: In: Catalysis IV. Hydrocarbon Synthesis, Hydrogenation, and Cyclization (P. H. Emmett, ed.), Reinhold, New York, Ch. 1–3 (1956).
Arrhenius, G. and Alfvén, H.: Earth Planet. Sci. Lett. 10, 253–267 (1971).
Arrhenius, G. and Raub, C. J.: J. Less-Common Metals 62, 417–430 (1978).
Asinger, F.: Paraffins, Chemistry and Technology, Pergamon Press, New York (1968).
Bandurski, E. L. and Nagy, B.: Geochim. Cosmochim. Acta 40, 1397–1406 (1976).
Bauer, H.: Naturwissenschaften 67, 1–6 (1980).
Belsky, T. and Kaplan, I. R.: Geochim. Cosmochim. Acta 34, 257–278 (1970).
Boato, G.: Geochim. Cosmochim. Acta 6, 209–220 (1954).
Brecher, A. and Arrhenius, G.: Nature Physical Science 230, 107–109 (1971).
Breger, I. A., Zubovic, P., Chandler, J. C. and Clarke, R. S.: Nature 236, 155–158 (1972).
Briggs, M. H.: Nature 197, 1290 (1963).
Broten, N. W., Oka, T., Avery, L. W., MacLeod, J. M. and Kroto, H. W.: Astrophys. J. 223, L105–L107 (1978).
Bunch, T. E. and Chang, S.: Geochim. Cosmochim. Acta 44, 1543–1577 (1980).
Cain, D. G., Weitkamp, A. W. and Bowman, N. J.: Ind. Eng. Chem. 45, 359 (1953).
Calvin, M.: Chemical Evolution, Clarendon Press, Oxford, Ch. 4–6 (1969).
Cameron, A. G. W.: Icarus 1, 13–69 (1962).
Cameron, A. G. W.: In: Interstellar Dust and Related Topics (J. M. Greenberg and H. C. Van de Hulst, eds.), D. Reidel, Dordrecht, p. 545–547 (1973).
Cameron, A. G. W.: In: Protostars and Planets (T. Gehrels, ed.) Univ. of Arizona Press, Tucson, 453–487 (1978).
Cameron, A. G. W. and Pine, M. R.: Icarus 18, 377–406 (1973).
Chang, S. and Mack, R.: Lunar Planet. Sci. IX, 157–159 (1978).
Churchwell, E., Winnewisser, G. and Walmsley, C. M.: Astron. Astrophys. 67, 139
Clayton, D. D.: Moon and Planets 19, 109–137 (1978).
Clayton, R. N.: Science 140, 192–193 (1963).
Clayton, R. N.: Ann. Rev. Nucl. Part. Sci. 28, 501–522 (1978).
Clegg, R. E. S.: Mon. Not. R. Astr. Soc. 191, 451–455 (1980).
Crabb, J., Lewis, R. S. and Anders, E.: Lunar Planet. Sci. XI, 174–176 (1980).
Cronin, J. R. and Moore, C. B.: Science 172, 1327–1329 (1971).
Cronin, J. R. and Moore, C. B.: Geochim. Cosmochim. Acta 40, 853 (1976).

Cronin, J. R., Pizzarello, S., and Moore, C. B.: Science 206, 335 (1979).

Dayhoff, M. O., Lippincott, E. R. and Eck, R. V.: Science 146, 1461–1464 (1964).

Delsemme, A. H.: In: Comets, Asteroids, Meteorites: Interrelations, Evolution, and Origins (A. H. Delsemme, ed.), Univ. of Toledo Press, 3–13 and 453–467 (1977).

DuFresne, E. R. and Anders, E.: Geochim. Cosmochim. Acta 26, 1085–1114 (1962).

Eberhardt, P., Jungck, M. H. A., Meier, F. O. and Niederer, F.: Astrophys. J. 234, L169–L171 (1979).

Eck, R. V., Lippincott, E. R., Dayhoff, M. O., and Pratt, Y. T.: Science 153, 628–633 (1966).

Folsome, C. E., Lawless, J. G., Romiez, M. and Ponnamperuma, C.: Geochim. Cosmochim. Acta 37, 455–465 (1973).

Freeman, C. G., Harland, P. W., and McEwan, M. J.: Astrophys. Lett. 19, 133–135 (1978).

Freeman, C. G., Harland, P. W., and McEwan, M. J.: Mon. Not. R. Astr. Soc. 187, 441 (1979).

Friedel, R. A. and Sharkey, A. G. Jr.: Science 139, 1203–1205 (1963).

Friedman, N., Bovee, H. H. and Miller, S. L.: J. Org. Chem. 35, 3230 (1970).

Friedman, N., Bovee, H. H. and Miller, S. L.: J. Org. Chem. 36, 2894–2897 (1971).

Galwey, A.: Nature 223, 1257–1260 (1969).

Galwey, A.: Geochim. Cosmochim. Acta 36, 1115–1130 (1972).

Gammon, R. H.: Chem. Engr. News 56, 21–34 (1978).

Geiss, J. and Reeves, H.: Astron. Astrophys., in press (1981).

Gelpi, E., Han, J., Nooner, D. W. and Oró, J.: Geochim. Cosmochim. Acta 34, 965–979 (1970).

Gelpi, E. and Oró, J.: Geochim. Cosmochim. Acta 34, 981–994 (1970a).

Gelpi, E. and Oró, J.: Geochim. Cosmochim. Acta 34, 995–1005 (1970b).

Grossman, L.: Geochim. Cosmochim. Acta 36, 597–619 (1972).

Grossman, L. and Clark, S. P., Jr.: Geochim. Cosmochim. Acta 37, 635–649 (1973).

Grossman, L. and Larimer, J. W.: Rev. Geophys. Space Phys. 32, 71–101 (1974).

Hayatsu, R.: Science 146, 1291–1293 (1964).

Hayatsu, R.: Science 149, 443–447 (1965).

Hayatsu, R., Studier, M. H., Oda, A., Fuse, K. and Anders, E.: Geochim. Cosmochim. Acta 32, 175–190 (1968).

Hayatsu, R., Studier, M. H. and Anders, E.: Geochim. Cosmochim. Acta 35, 939–951 (1971).

Hayatsu, R., Studier, M. H., Matsuoka, S. and Anders, E.: Geochim. Cosmochim. Acta 36, 555–571 (1972).

Hayatsu, R., Studier, M. H., Moore, L. P. and Anders, E.: Geochim. Cosmochim. Acta 39, 471–488 (1975).

Hayatsu, R., Matsuoka, S., Scott, R. G., Studier, M. H. and Anders, E.: Geochim. Cosmochim. Acta 41, 1325–1339 (1977).

Hayatsu, R., Winans, R. E., Scott, R. G., McBeth, R. L., Moore, L. P. and Studier, M. H.: Science 207, 1202–1204 (1980a).

Hayatsu, R., Scott, R. G., Studier, M. H., Lewis, R. S. and Anders, E.: Science 208, 1515–1518 (1980b).

Hayes, J. M.: Geochim. Cosmochim. Acta 31, 1395–1440 (1967).

Hayes, J. M. and Biemann, K.: Geochim. Cosmochim. Acta 32, 239–267 (1968).

Hayward, D. O. and Trapnell, B. M. W.: Chemisorption, Butterworths, London (1964).

Herbig, G. H.: Mém. Soc. Roy. Sci. Liège XIX, 13–26 (1970).

Herbst, E.: Astrophys. J. 205, 94 (1976).

Hodgson, G. W. and Ponnamperuma, C.: Proc. Natl. Acad. Sci. U.S. 59, 22–28 (1968).

Hodgson, G. W. and Baker, B. L.: Geochim. Cosmochim. Acta 33, 943–958 (1969).

Huntress, W. T. Jr.: Astrophys. J. Suppl. 33, 495 (1977).

Huntress, W. T. Jr. and Mitchell, G. F.: Astrophys. J. 231, 456–467 (1979).

Jordan, J., Kirsten, T. and Richter, H.: Z. Naturforsch. 35a, 145–170 (1980).

Jungclaus, G., Cronin, J. R., Moore, C. B. and Yuen, G. U.: Nature 261, 126 (1976).

Kerridge, J. F., Mackay, A. L. and Boynton, W. V.: Science 205, 395–397 (1979).

Khare, B. N. and Sagan, C.: In: Molecules in the Galactic Environment (M. Gordon and L. Snyder, eds.), J. Wiley & Sons, New York, 399–407 (1973).

Kölbel, H. and Trapper, J.: Angew. Chem. 78, 908–909 (1966).

Kolodny, Y., Kerridge, J. F. and Kaplan, I. R.: Earth Planet. Sci. Lett. 46, 149–158 (1980).

Kroto, H. W., Kirby, C., Walton, D. R. M., Avery, L. W., Broten, N. W., MacLeod, J. M. and Oka, T.: Astrophys. J. 219, L133–L137 (1978).

Kung, C.-C. and Clayton, R. N.: Earth Planet. Sci. Lett. 38, 421–435 (1978).

Kung, C.-C., Hayatsu, R., Studier, M. H. and Clayton, R. N.: Earth Planet. Sci. Lett. *46*, 141–146 (1979).

Kvenvolden, K. A.: Orig. of Life *5*, 71–86 (1974).

Kvenvolden, K. A., Lawless, J., Pering, K., Peterson, E., Flores, J., Ponnamperuma, C., Kaplan, I. R. and Moore, C.: Nature *228*, 923–926 (1970).

Kvenvolden, K. A., Lawless, J. G. and Ponnamperuma, C.: Proc. Natl. Acad. Sci. U.S. *68*, 486–490 (1971).

Lancet, M. S.: Ph. D. Thesis, Univ. of Chicago, 1972.

Lancet, M. S. and Anders, E.: Science *170*, 980–982 (1970).

Langer, W. D., Schloerb, F. P., Snell, R. L. and Young, J. S.: Astrophys. J. *239*, L125 (1980).

Larimer, J. W.: Geochim. Cosmochim. Acta *32*, 965–982 (1968).

Larimer, J. W.: Geochim. Cosmochim. Acta *37*, 1603–1623 (1973).

Larimer, J. W.: Geochim. Cosmochim. Acta *39*, 389–392 (1975).

Larimer, J. W. and Anders, E.: Geochim. Cosmochim. Acta *31*, 1239–1270 (1967).

Lawless, J. G.: Geochim. Cosmochim. Acta *37*, 2207–2212 (1973).

Lawless, J. G., Kvenvolden, K. A., Peterson, E., Ponnamperuma, C. and Moore, C.: Science *173*, 626–627 (1971).

Lawless, J. G., Kvenvolden, K. A., Peterson, E., Ponnamperuma, C.: Nature *236*, 66 (1972).

Lawless, J. G., Zeitman, B., Pereira, W. E., Simmons, R. E., and Duffield, A. M.: Nature *251*, 40–42 (1974).

Lawless, J. G. and Peterson, E.: Orig. of Life *6*, 3–8 (1975).

Lawless, J. G. and Yuen, G. U.: Nature *282*, 396 (1979).

Lederberg, J.: In: Biochemical Applications of Mass Spectrometry (G. R. Waller, ed.), J. Wiley & Sons, New York, 193–207 (1972).

Lederberg, J. and Cowie, D. B.: Science *127*, 1473–1475 (1958).

Lemmon, R. M.: Chem. Rev. *70*, 95–109 (1970).

Levin, B. Yu.: Russ. Chem. Rev. *38*, 65–78 (1969).

Levy, R. L., Wolf, C. J., Grayson, M. A., Gilbert, J., Gelpi, E., Updegrove, W. S., Zlatkis, A. and Oró, J.: Nature *227*, 148–150 (1970).

Levy, R. L., Grayson, M. A. and Wolf, C. J.: Geochim. Cosmochim. Acta *37*, 467–483 (1973).

Lewis, J. S.: Icarus *16*, 241–252 (1972).

Lewis, J. S. and Prinn, R. G.: Astrophys. J. *238*, 357–364 (1980).

Lewis, J. S., Barshay, S. S. and Noyes, B.: Icarus *37*, 190–206 (1979).

Lewis, R. S. and Anders, E.: Proc. Natl. Acad. Sci. U.S. *72*, 268–273 (1975).

Lewis, R. S., Gros, J. and Anders, E.: J. Geophys. Res. *82*, 779–792 (1977).

Lewis, R. S., Alaerts, L., Matsuda, J. and Anders, E.: Astrophys. J. *234*, L165–L168 (1979).

McCarthy, E. D. and Calvin, M.: Nature *216*, 642–647 (1967).

Meinschein, W. G.: Space Sci. Rev. *2*, 653–679 (1963).

Miller, S. L.: Science *117*, 528–529 (1953).

Miller, S. L.: J. Am. Chem. Soc. *77*, 2351 (1955).

Miller, S. L., Urey, H. C., and Oró, J.: J. Molec. Evol. *9*, 59–72 (1976).

Mitchell, G. F. and Huntress, W. T. Jr.: Nature *278*, 722–723 (1979).

Morris, M., Turner, B. E., Palmer, P. and Zuckerman, B.: Astrophys. J. *205*, 82–93 (1976).

Nagy, B.: Carbonaceous Meteorites, Elsevier, Amsterdam (1975).

Nagy, B., Meinschein, W. G. and Hennessy, D. J.: Annals N.Y. Acad. Sci. *108*, 534–552 (1963).

Nooner, D. W., Gibert, J. M., Gelpi, E., and Oró, J.: Geochim. Cosmochim. Acta *40*, 915–924 (1976).

Nooner, D. W. and Oró, J.: Geochim. Cosmochim. Acta *31*, 1359–1394 (1967).

Nooner, D. W. and Oró, J.: Adv. in Chem. Ser. *178* (E. L. Kugler & F. W. Steffgen, eds.), 159–171 (1979).

Norris, T. L.: Earth Planet. Sci. Lett. *47*, 43–50 (1980).

Onuma, N., Clayton, R. N. and Mayeda, T. K.: Geochim. Cosmochim. Acta *36*, 169–188 (1972).

Onuma, N., Clayton, R. N. and Mayeda, T. K.: Geochim. Cosmochim. Acta *38*, 189–191 (1974).

Oparin, A. I.: The Origin of Life, Dover, New York (1953).

Oró, J.: Space Life Sci. *3*, 507–550 (1972).

Oró, J. and Skewes, H. B.: Nature *207*, 1042–1045 (1965).

Oró, J. and Han, J.: Science *153*, 1393–1395 (1966).

Oró, J., Gelpi, E. and Nooner, D. W.: J. Brit. Interplanet. Soc. *21*, 83–98 (1968).
Oró, J., Gibert, J., Lichtenstein, H., Wikstrom, S. and Flory, D. A.: Nature *230*, 105–106 (1971a).
Oró, J., Nakaparksin, S., Lichtenstein, H. and Gil-Av, E.: Nature *230*, 107–108 (1971b).
Peltzer, E. T. and Bada, J. L.: Nature *272*, 443 (1978).
Penzias, A. A.: Astrophys. J. *228*, 430–434 (1979).
Pering, K. L. and Ponnamperuma, C.: Science *173*, 237–239 (1971).
Pichler, H., Schulz, H. and Kühne, D.: Brennstoff-Chem. *49*, 344 (1968).
Pikovskiy, Yu. I., Bashkirov, A. N. and Novak, F. I.: Dokl. Akad. Nauk. SSSR (Earth Sciences Section) *161*, 200–201 (1965).
Podosek, F. A.: Ann. Rev. Astron. Astrophys. *16*, 293–334 (1978).
Ponnamperuma, C., Lemmon, R. M., Mariner, R. and Calvin, M.: Proc. Natl. Acad. Sci. U.S. *49*, 737 (1963).
Ponnamperuma, C., Woeller, F., Flores, J., Romiez, M. and Allen, W.: Adv. in Chem. Ser. *80*, 280–288 (1969).
Porfir'ev, V. B.: Problem of the Inorganic Origin of Oil, Naukova Dumka, Kiev (1971).
Ring, D., Wolman, Y., Friedmann, N. and Miller, S. L.: Proc. Natl. Acad. Sci. U.S. *69*, 765–768 (1972).
Robert, F., Merlivat, L. and Javoy, M.: Nature *282*, 785–789 (1979).
Robert, F. and Epstein, S.: Meteoritics *15* (abstract), 355–356 (1980).
Robert, F., Becker, R. H. and Epstein, S.: Lunar Planet. Sci. *XI*, 935–937 (1980).
Robinson, R.: Nature *212*, 1291–1295 (1966).
Sagan, C. and Khare, B. N.: Science *173*, 417–420 (1971).
Sagan, C. and Khare, B. N.: Nature *277*, 102 (1979).
Schiff, H. I. and Bohme, D. K.: Astrophys. J. *232*, 740–746 (1979).
Seelig, H. S., Bowman, N. J. and Cady, W. E.: Ind. Eng. Chem. *45*, 343 (1953).
Shimoyama, A., Ponnamperuma, C. and Yanai, K.: Nature *282*, 394 (1979).
Smith, D. and Adams, N. G.: Astrophys. J. *220*, L87–L92 (1978).
Smith, J. W. and Kaplan, I. R.: Science *167*, 1367–1370 (1970).
Snell, R. L. and Wootten, A.: Astrophys. J. *228*, 748–754 (1979).
Srinivasan, B. and Anders, E.: Science *201*, 51–56 (1978).
Stephen-Sherwood, E. and Oró, J.: Space Life Sci. *4*, 5–31 (1973).
Stoks, P. G. and Schwartz, A. W.: Nature *282*, 709–710 (1979).
Stoks, P. G. and Schwartz, A. W.: Geochim. Cosmochim. Acta *44*, 563–569 (1981).
Studier, M. H., Hayatsu, R. and Anders, E.: Enrico Fermi Institute preprint No. 65–115, 19 pages (1965a).
Studier, M. H., Hayatsu, R. and Anders, E.: Science *149*, 1455–1459 (1965b).
Studier, M. H. and Hayatsu, R.: Anal. Chem. *40*, 1011–1013 (1968).
Studier, M. H., Hayatsu, R. and Anders, E.: Geochim. Cosmochim. Acta *32*, 151–174 (1968).
Studier, M. H., Hayatsu, R. and Anders, E.: Geochim. Cosmochim. Acta *36*, 189–215 (1972).
Studier, M. H., Hayatsu, R. and Winans, R. E.: In: Analytical Methods for Coal and Coal Products, Vol. *2* (C. Karr Jr., ed.), Academic Press, New York, 43 (1978).
Turekian, K. K. and Clark, S. P. Jr.: Earth Planet. Sci. Lett. *6*, 346–348 (1969).
Urey, H. C.: Proc. Natl. Acad. Sci. USA *38*, 351–363 (1952a).
Urey, H. C.: The Planets, Yale Univ. Press, New Haven (1952b).
Urey, H. C.: XII Intl. Congr. Pure & Applied Chem. (Plenary Lectures), 188–214 (1953), Almqvist & Wiksell, Stockholm.
Urey, H. C.: Quart, J. Roy. Astron. Soc. *8*, 23–47 (1967).
van der Velden, W. and Schwartz, A. W.: Geochim. Cosmochim. Acta *41*, 961 (1977).
Vdovykin, G. P.: Carbonaceous Matter of Meteorites (Organic Compounds, Diamonds, Graphite), Nauka Publishing Office, Moscow (1967); English translation, NASA TT F-582, Washington, D.C., (1970).
Vdovykin, G. P.: Usp. Sovrem. Biol. *87*, 49–60 (1979).
Walmsley, C. M., Winnewisser, G. and Toelle, F.: Astron. Astrophys. *81*, 245–250 (1980).
Watson, W. D.: Rev. Mod. Phys. *48*, 513 (1976).
Webster, A.: The Observatory *99*, 29–30 (1979).
Webster, A.: Mon. Not. Roy. Astr. Soc. *192*, 7P–9P (1980).
Whipple, F. L.: Proc. Natl. Acad. Sci. U.S. *52*, 565–594 (1964).

Whittaker, A. G.: Science *200*, 763–764 (1978).

Whittaker, A. G., Watts, E. J., Lewis, R. S. and Anders, E.: Science *208*, 1512–1514 (1980).

Winnewisser, G., Creswell, R. A. and Winnewisser, M.: Z. Naturforsch. *33a*, 1169–1172 (1978).

Wolman, Y., Haverland, W. J. and Miller, S.: Proc. Natl. Acad. Sci. U.S. *69*, 809–811 (1972).

Yang, C. C. and Oró, J.: In: Chemical Evolution and the Origin of Life (R. Buvet and C. Ponnamperuma, C., eds.), North-Holland, Amsterdam (1971).

Yoshino, D., Hayatsu, R. and Anders, E.: Geochim. Cosmochim. Acta *35*, 927–938 (1971).

Yuen, G. U. and Kvenvolden, K. A.: Nature *246*, 301–303 (1973).

Zeitman, B., Chang, S. and Lawless, J. G.: Nature *251*, 42–43 (1974).

Zuckerman, B., Turner, B. E., Johnson, D. R., Clark, F. O., Lovas, F. J., Fourikis, N., Palmer, P., Morris, M., Lilley, A. E., Ball, J. A., Gottlieb, C. A., Litvak, M. M. and Penfield, H.: Astrophys. J. *196*, L99–L102 (1975).

The Chemistry of Interstellar Molecules

Gisbert Winnewisser

I. Physikalisches Institut, Universität zu Köln, Köln, FRG

Table of Contents

1 Introduction

During the past decade, molecular line astronomy has contributed substantial new information about a variety of long standing astronomical problems such as (i) the distribution of matter within our galaxy and others, (ii) the mass loss associated with old and young stars, (iii) the formation of young stars occuring in dense molecular clouds and, (iv) the determination of isotope ratios as a useful indicator of the past chemical history of the Galaxy. Thus molecular line astronomy has opened several new areas of research and has widened our understanding of the cosmos. The most fascinating of these new research activities certainly seems to be cosmochemistry, the chemistry prevailing in the highly diluted medium of space as compared to terrestrial conditions. This is for the simple reason that we all share a certain curiosity in understanding of (i) how the cosmically most abundant elements H, C, N, O (with the exception of He) bind together under the extreme conditions of interstellar space to form simple and fairly complex molecules and (ii) how these rather recent interstellar results relate to and influence our knowledge of the origin and early history of the solar system.

Although some of the organic interstellar molecules are by far not as complex as biologically important molecules such as deoxyribonucleic acid (DNA) or ribonucleic acid (RNA), which govern the process of reproduction, the most basic property of life, they are however fairly exotic even when compared with present day organic chemistry since not all interstellar molecules have been synthesized in the laboratory. Among the known interstellar molecules, one finds all the smaller molecules or functional groups out of which larger molecules and eventually the biologically complex molecules can form. All the ingredients are produced there in the hostile environment of interstellar space, awaiting conditions which allow the formation of even biologically important molecules. Whether these conditions are ever met by interstellar space seems unlikely, or whether they are only found in atmospheres of certain planets in the close neighbourhood of stars remains an intriguing question.

On the other hand one finds interstellar molecules such as water, H_2O, methyl- or ethylalcohol CH_3OH, CH_3CH_2OH respectively which are well known to us from common day life.

The aim of this article is to give a short outline of current theories of molecule formation and destruction in interstellar clouds, together with a short summary of the observational material which has been accumulated up to early 1981. Although this article will address itself predominantly to simple molecules a section on complex molecules has been added. We will, therefore, discuss some general aspects of cosmochemistry and then turn to molecule formation in diffuse clouds followed by a discussion of the chemistry of dense interstellar clouds. A section has been added to summarize recent observational results and theoretical proposals in understanding the formation of intermediate and complex molecules, an area of considerable current activity. Finally the article closes with a short summary of the molecular species found in planetary atmospheres and a short discussion of what the relation might be to the interstellar molecules.

The subject of molecule formation and destruction has been reviewed by several authors (e.g. Herbst and Klemperer 1976, Watson 1976, Herbst 1978 and Watson

1978). In this cosmochemistry series "Topics in Current Chemistry", Winnewisser, Mezger and Breuer 1974 have given a general review of interstellar molecules, with some consideration of molecule formation mechanisms. Recently Winnewisser, Churchwell and Walmsley 1979 have given a detailed account of the "Astrophysics of Interstellar Molecules" with a chapter specially devoted to molecule formation mechanisms. This article is based on these earlier reviews with emphasis on some of the more recent developments.

We will not mention effects on molecular formation due to shocks and shock fronts in dense molecular clouds, nor will we discuss the chemistry of the circumstellar environment, where an abundance of molecular species has been detected during the past several years. In the warm, dense envelopes of stars the abundances can be matched by chemical-equilibrium calculations, in contrast to the chemical reactions which can take place in the cold interstellar molecular clouds. For example theoretical calculations based on chemical equilibrium have been performed for the expanding molecular envelope of the cool carbon star IRC +10216 by McCabe et al. (1980), in agreement with the observed molecular column-densities.

2 Astrophysical Scenario

Our Galaxy consists of stars and matter which has either not yet participated in the formation of stars or has already been expelled by them, i.e. interstellar matter is probably composed of original material and the "ashes" from the nuclear burning processes returned by stars in their final stages of evolution. This matter constitutes together with the radiation field covering all wavelengths from γ-rays to the metre-wavelength radio background, the interstellar medium. In the Galaxy interstellar matter comprises presently 10% of the total galactic mass which is estimated to be about 1.1×10^{11} M$_\odot$, with one solar mass 1 M$_\odot \sim 2 \times 10^{33}$ g.

The dawn of the galaxy probably started with its entire mass in the form of hydrogen gas (and possibly helium), 90% of which has since been converted into stars. The remaining mass comprises now the matter between the stars. Interstellar matter proper consists of gas and dust which are heterogeneously dispersed between the stars. It is now known from the observational data that the distribution of interstellar matter on a galactic scale always follows closely that of the stars indicating their intimate connection: stars do form in regions of high concentration of interstellar matter. While stars are concentrated in a thin flat disk about 500 to 600 pc in width (where 1 pc = 1 par sec = 3.26 light years = 3.09×10^{18} cm) and about 40 kpc in diameter, interstellar matter is even more closely confined to the galactic plane. Within the galactic disk interstellar matter is not distributed uniformly, it is very clumpy and patchy giving rise to the interstellar clouds.

The principal constituents of interstellar matter are fine dust particles (with a diameter of several µm) and gas. One therefore believes that the dust-to-gas ratio is about constant throughout the galaxy with the possible exception of the galactic centre. Table 1 which is adopted from Winnewisser et al. (1979) summarizes

41

the various components of interstellar matter and also indicates the methods by which they can be investigated.

With practically no exception, observational data indicate that interstellar dust and gas always appear simultaneously. The extinction of starlight i.e. the combined effect of absorption and scattering at optical wavelength is still the main evidence for the existence of interstellar dust, although observations of infrared emission from heated dust grains near young stars furnish us with independent but strong evidence for its wide distribution. Their chemical composition is presently far from being settled, but there is now strong evidence that silicate and graphite are the main constituents with ice and traces of heavier elements surrounded by envelopes containing carbon nitrogen and oxygen. There are several fine reviews of the physical and chemical properties of interstellar grains (Aanestad and Purcell 1973) and the observational data which pertains to them (Savage and Mathis, 1979). It has however to be noted that although the interstellar dust particles account for only 1 % of the interstellar matter, they seem to play a vital but still uncertain role in the formation of interstellar molecules. In fact ever since the first attempts to understand the mechanisms which govern interstellar cloud chemistry, the question has arisen whether molecules are formed through gas-phase reaction sequences or by the alternative process, i.e. formation of molecules on grain surfaces. Although in this review there will be no attempt made to settle this issue, results are described which have some bearing on this question. To assess the intrinsic difficulties encountered in this fundamental issue of molecule formation mechanisms one should bear in mind that the strongest evidence for the importance of grain chemistry is the existence of H_2 the most abundant molecule in space, whereas the existence of HD furnishes us with the strongest evidence for the importance of gas phase chemistry. In addition, interstellar grains are also important because they shield the molecules from dissociation by UV radiation. Thus interstellar dust and gas seem to be connected like the master and his slave, but the question remains which of the two assumes the role of the master.

Table 1 indicates that the interstellar gas consists essentially of two components: the ionized and the neutral form. It is the investigation of the neutral component which is of special interest in an attempt to understand the general process of

Table 1. Interstellar matter (Winnewisser et al., (1979))

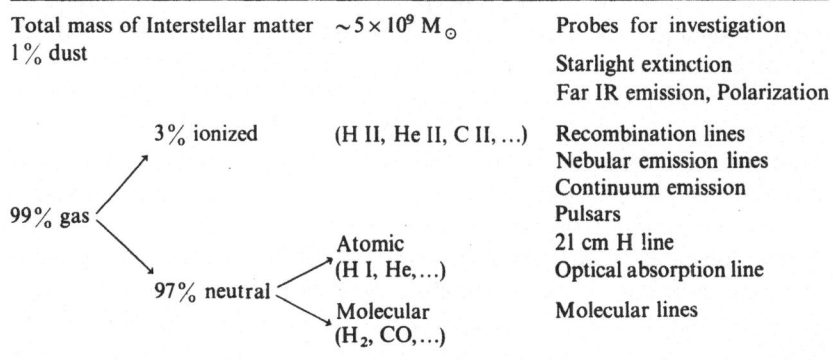

Total mass of Interstellar matter $\sim 5 \times 10^9$ M$_\odot$ 1 % dust		Probes for investigation
		Starlight extinction
		Far IR emission, Polarization
99 % gas	3 % ionized (H II, He II, C II, ...)	Recombination lines
		Nebular emission lines
		Continuum emission
		Pulsars
	97 % neutral — Atomic (H I, He, ...)	21 cm H line
		Optical absorption line
	Molecular (H_2, CO,...)	Molecular lines

molecular formation and reactions in interstellar space. One estimates that about 50% of the neutral component is in molecular form, with molecular hydrogen H_2 being the most abundant interstellar molecule followed by carbon monoxide, CO, as has become evident from recent large-scale CO surveys of the Galaxy. Estimates of the galactic mass of H_2 and CO can now be given with reasonably accuracy $M(H_2) = 2 \times 10^9\ M_\odot$, $M(CO) = 2 \times 10^6\ M_\odot$.

Although the average density of the interstellar gas within the Galaxy is about one hydrogen atom per cubic centimetre, it does vary considerably with galactic radius (see for example Winnewisser et al. 1979). However, the gas density is not uniformly distributed on the contrary it shows local concentrations of gas (and dust) with gas densities as high as 10^6 to $10^8\ cm^{-3}$. It is in these regions of higher than average gas density, called interstellar clouds where many different molecules are currently being found. Although by now a large variety of molecular clouds has been detected in interstellar space, there exists nothing like a standard molecular cloud. However, it has been widely accepted that one can describe the gas component as composed of three different phases: (i) the diffuse interstellar clouds with an average density of about 1 to 100 particles cm^{-3} and a temperature of $\lesssim 100$ K. (ii) the dense molecular clouds (density $\sim 10^3$–$10^6\ cm^{-3}$, all hydrogen is essentially in molecular form) and (iii) a very diluted (<0.1 particle cm^{-3}) but hot (10^3 K $\lesssim$ T $\lesssim 10^4$ K) intercloud medium. Only simple molecules are found in the diffuse interstellar clouds whereas the bulk of the molecular detections pertain to the dense molecular clouds. From the molecular point of view the intercloud medium is of little importance and will not be discussed further.

The dense molecular clouds span a wide scale of physical parameters such as size (1 — 200 pc) density ($10^3 \lesssim n_{H_2} \lesssim 10^8\ cm^{-3}$), temperature $5 \lesssim T_{kin} \lesssim 150$ and total masses between $10 \lesssim M_{cloud} \lesssim 10^6\ M_\odot$. Recently Winnewisser et al. 1979 have attempted a cloud classification on the basis of the observed molecular line-widths. This classification is reproduced in Table 2. Since the line-width carries a large amount of intrinsic information about the physical properties of the clouds, from which the lines emanate, this classification also serves as a good indicator of the chemical reaction schemes assumed to prevail in the individual cloud types. Thus we assume that certain reaction schemes are associated with specific types of molecular clouds.

We will therefore consider in the following sections some basic ideas and observations of the chemistry pertaining to diffuse interstellar clouds and cold dark and dense clouds.

3 General Aspects of Interstellar Chemistry

Since the first attempts to cast the subject of interstellar chemistry into quantitative terms by Bates and Spitzer (1951) in understanding the abundance of CH and CH^+, there has been considerable discussion about the relative importance of gas phase chemistry versus catalytic processes on interstellar grain surfaces. This issue has remained the great unknown in interstellar chemistry and has assumed new importance by being shifted to the question to which extent both mechanisms contribute to the formation of larger organic molecules. Ever since the early success in under-

Table 2. Interstellar molecules detected by optical techniques in diffuse clouds (Winnewisser et al., (1979))

Molecule	Transition	Wavelength (Å)	Other transitions	Column densities (cm^{-2})	Source	Year of discovery
^{12}CH	$A^2\Delta - X^2\Pi(0,0)\ R_2(1)$	4300	Y	$\sim 10^{13}$	~ 40	1937
	$B^2\Sigma^- - X^2\Pi(0,0)\ ^PQ_{12}(1)$	3890	Y			1941
	$C^2\Sigma^+ - X^2\Pi(0,0)\ ^PQ_{12}(1)$	3146	Y			1960
^{12}CH$^+$	$A^1\Pi - X^1\Sigma(0,0)\ R(0)$	4233	Y	10^{13}	~ 60	1937
^{13}CH$^+$	$A^1\Pi - X^1\Sigma(0,0)\ R(0)$	4232	Y	10^{13}	~ 2	1969
CN	$B^2\Sigma^+ - X^2\Sigma^+(0,0)\ R(0)$	3875	Y	10^{12}	14	1938/39
CO	$A^1\Pi - X^1\Sigma(1,0)$	1510	Y	10^{15}	ζ Oph	1971
	$C^1\Sigma^+ - X^1\Sigma^+(0,0)\ R(0)$	1088	Y	$\sim 10^{13}$	3	1973
	$E^1\Pi - X^1\Sigma^+(0,0)\ R(0)$	1076	Y	$\sim 10^{13}$	3	1973
^{13}CO	$A^1\Pi - X^1\Sigma(2,0)$	1476	Y	$\sim 10^{13}$	ζ Oph	1971
H$_2$	$B^1\Sigma_u - X^1\Sigma_g(0,0)$	1108	Y	10^{20}	15	1970
HD	$B^1\Sigma_u^+ - X^1\Sigma_g^+(3,0)\ R(0)$	1066	Y	$\sim 10^{14}$	9	1973
OH	$A^2\Sigma^+ - X^2\Pi$	3078			2	1976
	$D^2\Sigma - X^2\Pi(0,0)\ Q_1(3/2)$	1222	Y	$\sim 10^{14}$	ζ Oph	1976
C$_2$	$A^1\Pi_u - X^1\Sigma_g^+(2,0)\ R(0)$	8758	Y	$\sim 10^{13}$	ζ PER	1980

standing the formation of molecular hydrogen by association on the surfaces of grains (Hollenbach and Salpeter 1971, Hollenbach, Werner and Salpeter, 1971), surface chemistry has been hampered seriously by the lack of predictability. This is connected with the intrinsic difficulties of the process itself. Firstly there is still now a great lack of detailed knowledge of what precisely grains are made of, and secondly the details of the catalytic processes remain rather uncertain and speculative. Although there are essentially three steps involved in the molecular formation on grain surfaces[1], (each associated with is own uncertainty) it is the final step which poses the central problem. With the exception of H_2 it has not been shown convincingly for other molecular species how the product molecule could acquire sufficient kinetic energy to escape from the grain surface. A summary of reactions on the surfaces of dust grains is given by Watson, 1978.

On the other hand gas-phase reactions between positive ions and neutrals have been recognized in the early 70's to be a basic process giving birth to interstellar molecules (Herbst and Klemperer, 1973, 1976). No activation energies and exothermicity for most of these reactions produces appreciable reaction rates even at the low temperatures $(T \lesssim 100 \text{ K})$ of interstellar clouds. It therefore gives them preference over neutral-neutral reactions, which are usually hampered by problems of activation energy. Negative-ion neutral reactions suffer from the fact that for most cases they are endothermic and are therefore of limited signifiance to astrophysical problems. Thus based on the idea of positive-ion chemistry it has become possible to predict quantitatively the abundances of simple and some intermediate size interstellar molecules. With laboratory measured and/or calculated and estimated reaction rates, the results of positive-ion gas phase chemistry in the diluted medium of interstellar space have become predictable, a commodity which has certainly contributed to the success of ion molecule reactions. Based on this concept the early prediction of the formation of HCO^+ and N_2H^+ were key-successes for the ion-molecule scheme. The properties of such reactions have been discussed by McDaniels et al. 1970, and their astrophysical significance by Dalgarno and Black 1976. Huntress 1976 discusses and summarises recent laboratory measurements.

In summary we may then note concerning the question of the relative importance of catalytic processes on grain surfaces versus gas phase reactions that the results for simple molecules lend a certain preference to ion molecule reactions over the competing surface chemistry models. Strong evidence for catalytic surface reactions comes from the formation of molecular hydrogen. Recombination of hydrogen atoms on grain surfaces according to the reaction

$$H + H + \text{grain} \rightarrow H_2 + \text{grain}$$

with a formation rate $(\text{cm}^{-3} \text{ sec}^{-1})$ $k_1 n$ (HI) n where $n = n$ (HI) $+ 2n(H_2)$ and $k_1 \sim 10^{-17} \text{ cm}^3 \text{ sec}^{-1}$ is reasonably well understood and will not be repeated here although there are still problems associated with it (see for example recent summaries: Winnewisser et al. 1979, Hollenbach and McKee, 1980). For instance,

---

[1] In short the three steps are: the sticking ability of a molecule to the surface, subsequent mobility on the grain surface to encounter a reaction partner and finally reaction and ejection from the surface.

the initial rotational and vibrational distribution of newly created H_2 molecules cannot be predicted with certainty (Hunter and Watson 1978) which expresses itself in the observed rotational distribution of H_2 and affects the associated analysis and the calculation of the heating rate.

For the case of high temperatures found in the vicinity of hot stars, or in cloud regions under the influence of a shock wave, or in ionized nebulae, several authors have suggested (c.f. Jura, 1975, Dalgarno and McCray 1973) a gas phase formation scheme which could be a significant source for molecular hydrogen, H_2. The associative detachment reaction

$$H + H^- \rightarrow H_2 + e$$

has a measured rate constant of 1.3×10^{-9} cm^3 sec^{-1} (Schmeltekopf et al. 1967). From the formation rate of H^- via the reaction

$$H + e \rightarrow H^- + h\nu; \qquad k = 5 \times 10^{-17} \text{ cm}^3 \text{ sec}^{-1}$$

and the estimated lifetime of H^- against photodestruction (5×10^6 sec) one can deduce that this gas phase mechanism is likely to be less effective by two orders of magnitude than surface reactions in areas with a ratio of electron density n_e to hydrogen density $n = n(HI) + 2n(H_2)$ $n_e/n < 10^{-3}$, which is typical for interstellar clouds.

Although it would really be surprising if H_2 were to remain the only molecule to be produced on grain surfaces, it has emerged that the formation of other simple molecules can now be explained satisfactorily — not without problems though — by gas phase reactions. Evidence but not proof for positive-ion gas-phase reactions comes from several rather independent areas (c.f. Watson, 1980)

(i) the formation of HD and OH in diffuse clouds

(ii) abundance predictions for HCO^+ and N_2H^+ in dense clouds

(iii) the nearly equal interstellar abundance ratio of HCN and HNC

(iv) deuterium enchancement in the deuterium to hydrogen ratio in molecules

(v) the apparent fractionation of carbon isotopes observed in several clouds for various molecules

(vi) the agreement between prediction and deduced electron densities in dense clouds.

The next chapter will present a short outline of the basic ideas which have gone into the model of gas-phase chemistry centered around positive-ion reactions and we will illustrate the reaction paths of some selected simple interstellar molecules. In this case one has to consider two limiting cases of interstellar sources, (i) the diffuse interstellar cloud and (ii) the dense molecular clouds which themselves will have to be subdivided into several different cloud types.

4 Chemistry in Molecular Clouds

The primary energy source behind the ion-molecule scheme in interstellar molecular clouds is the cosmic ray ionization of H, H_2 and He, which can be transferred efficiently to less abundant atoms or species, notably C, N and O. Thus their effective time scale for ionization is reduced by a factor proportional to their abundance. In addition, exothermic reactions between positive ions and neutrals occur with no activation energy and with rate coefficients near 10^{-9} cm^3 sec^{-1} at room temperature. Since these rates usually do not change with temperature (some increase at low temperatures) they remain fast at low temperatures. These general characteristics make ion-molecule reactions important for interstellar cloud chemistry, i.e. for the diffuse interstellar clouds as well as for the cool dense dark molecular clouds.

4.1 Diffuse Interstellar Clouds

The chemical composition of diffuse interstellar clouds is simple and essentially limited to diatomic molecules which have been summarized in Table 2, together with the wavelength region where detection was made. It has to be noted, however, that there is a marginal detection of H$_2$O at about the 2σ level by Snow (1980). If confirmed a new H$_2$O mechanism has to be thought of.

Diffuse clouds are tenuous concentrations of interstellar gas and dust that do not block entirely the light of stars which are located behind them. They can be studied by absorption spectroscopy and as seen from Table 2, they were already studied as early as 1940. Although diffuse clouds are chemically simpler than are dense molecular clouds the assessment of the formation and destruction mechanisms has its own difficulties associated with it. Just because of the lower density, photoionization and photodissociation processes play a significant role in altering the otherwise simple chemistry of the diffuse clouds. The formation of HD may serve as a standard example.

Cosmic ray ionization of H leads to the formation of HD through a sequence of reactions. The resonant charge-exchange reaction, whose rate constant has been given (Watson et al. 1978)

$$H^+ + D \rightarrow D^+ + H \qquad (\sim 10^{-9} \text{ cm}^3 \text{ sec}^{-1})$$

furnishes an important source of D$^+$, which reacts by the gas-phase reaction

$$D^+ + H_2 \rightarrow HD + H^+ \qquad (\sim 10^{-9} \text{ cm}^3 \text{ sec}^{-1})$$

whose rate constant has been measured by Fehsenfeld et al. 1973. The observed column density ratio HD/H$_2$ is 10^{-6} with about an order of magnitude variation about this value. The D/H abundance ratio has been measured for a distance up to 200 pc from the sun directly in the ultraviolet and is 10^{-5} (York and Rogerson 1976). It may be noted that the HD abundance is directly proportional to the cosmic ray flux and the cosmic D/H ratio. The latter ratio can be determined once the cosmic ray flux is known. This has been done from the OH abundance via the charge exchange reaction

$$H^+ + O \rightarrow O^+ + H$$

followed by

$$O^+ + H_2 \rightarrow OH^+ + H$$

47

which transfers the ionization to a molecule that is subject to rapid neutralization (Watson 1973). For ζ Oph, ζ Per, and o Per the D/H ratio has been determined to be 4×10^{-5} (Hartquist, Black and Dalgarno, 1978), in agreement with the value from York and Rogerson.

Other important reactions take place between C^+ and H_2. Black and Dalgarno 1973 suggested that radiative association

$$C^+ + H_2 \rightarrow CH_2^+ + h\nu$$

can occur and can trigger an entire reaction sequence with the end products CH, CH^+, C_2, C_2H and $C_2H_2^+$. The latter two species are important species for the formation of complex carbon chain molecules in dense clouds as will be discussed in the next section. CN and molecular carbon C_2 are other important species. While CN is produced by various reaction sequences involving atomic nitrogen ($CH^+ + N \rightarrow CN + H^+$, or $C_2^+ + N \rightarrow CN + C^+$) and the CH cycle, C_2 is produced by reaction of C^+ with the CH cycle ($CH + C^+ \rightarrow C_2^+ + H$) (Black and Dalgarno 1977 and Black et al. 1979). The observational value of the column density of 1.4×10^{13} cm^{-2} towards ζ Per (Chaffee Jr. et al. 1980) is in remarkable good agreement with the predicted value from the model by Black et al. 1979. Table 3 which gives a summary of predicted and observed column densities for five molecular species has been taken from Dalgarno 1980. It is noted that there is generally good agreement between theoretical prediction and observational results with the exception of CH^+. The chemical model produces too little CH^+. Dalgarno 1980 discusses several possibilities all of which encounter difficulties, such as that the rate constant of $C^+ + H_2 \rightarrow CH_2^+ + h\nu$ is too low or the destruction mechanisms are not taken into account properly.

It is also of interest to note that the chlorine chemistry in diffuse clouds has been discussed (Dalgarno et al. 1974, and Black and Dalgarno, 1977) showing that Cl^+ will react with H_2:

$$Cl^+ + H_2 \rightarrow HCl^+ + H$$
$$HCl^+ + H_2 \rightarrow H_2Cl^+ + H$$
$$H_2Cl^+ + e \rightarrow HCl + H$$

with the destruction of hydrogen chloride by either photodissociation or with C^+

$$HCl + C^+ \rightarrow CCl^+ + H \, .$$

Table 3. Column densities log N (cm^{-2}) (Dalgarno, 1980)

	ζ Oph		ζ Per		o Per	
	Obs	Th	Obs	Th	Obs	Th
CO	15.0–15.2	15.0	14.7–15.0	15.0	14.7–15.0	15.1
CH	13.5–13.6	13.6	13.0–13.4	13.6	13.4–13.6	13.5
CN	12.94	12.95	12.6	12.5	12.3–12.7	11.9
C_2	12.9	12.9	13.1	13.1	—	13.3
CH^+	13.0	11.4	12.2	11.5	12.7	11.5

Furthermore several authors point out that for a hot environment gas phase chemistry is altered drastically in that endothermic reactions with H_2 become progressively more important (Iglesias and Silk 1978; Elitzur nach De Jong 1978, Hartqvist, Oppenheimer and Dalgarno 1980), and molecular ions such as SiH^+, SH^+, NaH^+, MgH^+, FeH^+ and HCl^+ may become detectable.

4.2 Dense Molecular Clouds

Dense molecular clouds, often also called dark clouds, block entirely the light of stars which lie behind them, and can therefore be studied observationally only by radio astronomy or infrared techniques. These clouds have a visual extinction in excess of $A_v \gtrsim 10$ which corresponds to a gas density of $n_{H2} \sim 10^4 \, cm^{-3}$ and a kinetic temperature usually well below $T \sim 100$ K, typically between 10 and 25 K. Within the last ten years, the investigation of these dark molecular clouds has become almost entirely the domain of radio astronomy although now the first very promising results by infrared astronomy reveal the power of this new branch of spectroscopy.

The dark molecular clouds are chemically very rich in comparison to the diffuse clouds. They harbor all of the presently observed more than 50 chemical species which are summarized in Table 4. Clearly dark clouds are a very heterogenous group which have been classified by Winnewisser, Churchwell and Walmsley 1979. We will reproduce this classification here which is based on observed physical and astronomical parameters. Table 5 groups the dark clouds into four different classes according to observed molecular line-widths, size of the cloud and other astronomical parameters with which the cloud is associated. The first class are the giant molecular clouds, those clouds which are associated with ionized regions of gas, H^+ regions (or often referred to as H II regions). The large molecular line width is a typical characteristic of these clouds. Together with the dark dust clouds which form the second category they are the clouds in which star formation is occuring and probably only there. The third and fourth group are intrinsically connected with the immediate neighbourhood of stars. The protostellar environment is associated with the formation of young stars whereas the envelopes of late type stars describes the molecular clouds which can form around old stars which are losing processed mass to the interstellar medium. Although we will not discuss the chemistry of the protostellar environment or the chemistry of late type stars, it is of interest to note here that the abundance ratios of the molecular species (also isotopic) in circumstellar envelopes can be quite different from that of the interstellar medium. For example in IRC +10216 the observed circumstellar (HCN)/(HNC) ratio (~ 100) is very much different from the observed interstellar ratio (~ 1), where both species are about equally abundant. The circumstellar ratio is in agreement with calculated values assuming chemical equilibrium at ~ 1000 K. A summary on the chemical composition and physical properties of the envelopes of late type stars is given by Zuckerman 1980.

In the following two sections the discussion will be confined to the chemistry of dense molecular clouds, where one can assume that effects of photodissoziation and photoionization are unimportant, quite in contrast to diffuse clouds. Furthermore, in dense clouds hydrogen is predominantly in the molecular form, H_2, and reactions

Table 4. Interstellar Molecules

2	3	4	5	6	7	8	9	10	11
H_2	H_2O								
OH	H_2S								
SO	N_2H^+	NH_3							
SiO	SO_2								
SiS	HNO								
NO	O_3								
NS									
CH^+	HCN	H_2CO	HC_3N	CH_3OH	HC_5N	$HCOOCH_3$	HC_7N	—	HC_9N
CH	HNC	$HNCO$	C_4H	CH_3CN	CH_3CCH		$(CH_3)_2O$		
CN	C_2H	H_2CS	H_2CNH	CH_3SH	CH_3NH_2		CH_3CH_2OH		
CO	HCO	$HNCS$	NH_2CN	NH_2CHO	CH_3CHO		CH_3CH_2CN		
CS	HCO^+	C_3N	$HCOOH$		H_2CCHCN				
	OCS		H_2C_2O						
	HCS^+								

Isotopically substituted molecules

By optical techniques

H_2	H_2, HD
CH^+	$^{12}CH^+$, $^{13}CH^+$
CO	$^{12}C^{16}O$, $^{13}C^{16}O$
H_2O	$H_2^{16}O$, $HD^{16}O$, $H_2^{18}O$
N_2H^+	$^{14}N^{14}NH^+$, $^{14}N^{14}ND^+$
HCO^+	$H^{12}C^{16}O^+$, $H^{13}CO^+$, $D^{12}C^{16}O^+$, $H^{12}C^{18}O$
HCN	$H^{12}C^{14}N$, $H^{13}C^{14}N$, $H^{12}C^{15}N$, $D^{12}C^{14}N$
HNC	$H^{14}N^{12}C$, $H^{14}N^{13}C$, $D^{14}N^{12}C$

By radio techniques

OH	^{16}OH, ^{18}OH, ^{17}OH
SiO	$^{28}Si^{16}O$, $^{29}Si^{16}O$, $^{30}Si^{16}O$
CO	$^{12}C^{16}O$, $^{13}C^{16}O$, $^{12}C^{18}O$, $^{12}C^{17}O$, $^{13}C^{18}O$
CS	$^{12}C^{32}S$, $^{13}C^{32}S$, $^{12}C^{34}S$, $^{12}C^{33}S$
NH_3	$^{14}NH_3$, $^{15}NH_3$
H_2CO	$H_2^{12}C^{16}O$, $H_2^{13}C^{16}O$, $H_2^{12}C^{18}O$, $HDCO$
HC_3N	$H^{12}C^{12}C^{12}C^{14}N$, $H^{13}C^{12}C^{12}C^{14}N$, $H^{12}C^{13}C^{12}C^{12}C^{14}N$, $H^{12}C^{12}C^{13}C^{14}N$, $DCCCN$
CH_3OH	$^{12}CH_3^{16}OH$, $^{13}CH_3OH$, CH_3OD

Table 5. Classification of molecular line sources (Winnewisser et al. 1979)

Class	Linear size (pc)	Typical linewidths (km s^{-1})^a	Density (cm^{-3})	Temperature (K)	Mass (M$_\odot$)
Molecular clouds associated with H II regions (giant)	1–50	3–30	10^4–10^6	20– 80	10^5–10^6
Dark dust clouds	1–10	1	10^3–10^5	10– 20	10^2–10^4
Envelopes of late type stars	0.01–0.5	25	10^4–10^6	100–1000	$\sim 10^{-2}$
Masers in a "proto-stellar" environment	10^{-5}–10^{-3}	0.1–2	$> 10^5$	100–1000	$\lesssim 10$

^{a} From the Doppler relation $\Delta v/v = -\Delta \upsilon/c$, 1 km s^{-1} corresponds to a linewidth (kHz) of 3.336v (GHz).

between molecular species are of importance particularly for the formation of larger molecules. In addition, physical conditions in dense clouds are considerably more complicated than in diffuse clouds and thus the observational data are often difficult to assess precisely. In particular, there are always problems associated with the conversion of observed line intensities to molecular abundances, numbers which are of primary interest to the chemist. The observational data have recently been reviewed by Winnewisser et al. 1979. The proceedings of the recent IAU Symposium No. 87 on Interstellar Molecules (B. H. Andrew, Ed. 1980) is an excellant compendium to the current observational and theoretical knowledge, the success and failures of astrophysics and astrochemistry and in understanding interstellar molecules.

4.3 The Ion-Molecule Scheme: HCO$^+$, N$_2$H$^+$, H$_2$O and NH$_3$

According to the ion-molecule scheme, the chemistry in dense molecular clouds is driven by the assumed cosmic ray ionization of the most abundant species: i.e. H, H$_2$ and He. The cosmic ray ionization rate ξ_{CR} should be 10^{-17} sec^{-1} based upon the cosmic ray flux measured at earth.

Ionization of the main gas component H$_2$, the most important basic process, leads essentially to H$_3^+$ which then either recombines or reacts further with molecules or molecular fragments. Thus the latter destruction mechanism of H$_3^+$ is of fundamental importance to ion chemistry in dense clouds by transferring the ionization to molecular or less abundant elements.

A likely sequence is

$$\text{H}_2 + \text{C.R.} \xrightarrow{98\%} \text{H}_2^+ + e + \text{C.R.}$$
$$\xrightarrow{2\%} \text{H}^+ + \text{H} + e + \text{C.R.}$$

This reaction is followed by

$$\text{H}_2^+ + \text{H}_2 \rightarrow \text{H}_3^+ + \text{H}$$

There are several pathways for the destruction of H$_3^+$.

Dissosiative recombination with an electron

$$H_3^+ + e \rightarrow H_2 + H$$

competes with reactions of the form

$$H_3^+ + A \rightarrow AH^+ + H_2$$

and

$$H_3^+ + AB \rightarrow HAB^+ + H_2$$

Carbon monoxide is known to be next to H_2 one of the most abundant molecules. In addition, it is highly probable that in dense molecular clouds most of the gas-phase carbon is in the form of CO (there could also be large amounts of C_2). It has been suggested that the reaction

$$H_3^+ + CO \rightarrow HCO^+ + H_2$$

should produce a large fraction of the interstellar HCO^+ through the destruction of H_3^+. Similarly one would expect

$$H_3^+ + N_2 \rightarrow N_2H^+ + H_2$$

and

$$H_3^+ + C_2 \rightarrow C_2H^+ + H_2$$

The prediction and identification of HCO^+ and N_2H^+ as key molecules of the positive-ion gas-phase chemistry paved the way to its general acceptance.

Interesting examples of the formation of intermediate molecules are illustrated by the possible exothermic reactions of H_3^+ with O and N and the reaction paths leading finally to H_2O and NH_3:

$$H_3^+ + O \rightarrow OH^+ + H_2$$
$$OH^+ + H_2 \rightarrow OH_2^+ + H$$
$$H_2O^+ + H_2 \rightarrow H_3O^+ + H$$

$$H_3O^+ + e \nearrow^{H_2O + H}_{\searrow OH + H_2}$$

Support for these reactions comes from laboratory measurements which show that their reaction rates are fast. Yet the interstellar importance of these reactions is not completely clear since the reaction of H_3^+ with atomic oxygen has to compete with the dissociative electron recombination of H_3^+.

Similarly, the reaction with atomic nitrogen produces NH_3, NH_2 and NH:

$$H_3^+ + N \rightarrow NH^+ + H_2$$
$$NH^+ + H_2 \rightarrow NH_2^+ + H$$
$$NH_2^+ + H_2 \rightarrow NH_3^+ + H \; .$$

The reaction

$$NH_3^+ + H_2 \rightarrow NH_4^+ + H$$

is important in spite of the fact that the rate constant is small ($\sim 10^{-13}$ cm^3 sec^{-1}). NH_3 can also be formed by either charge exchange reactions with heavy atoms of low ionization potential, i.e. Fe, He, Ca, Na according to:

$$NH_3^+ + Mg, ... \rightarrow NH_3 + Mg^+, ...$$

or by the dissociative electron recombination reaction of NH_5^+:

$$NH_3^+ + H_2 + h\nu \rightarrow NH_5^+$$
$$NH_5^+ + e \rightarrow NH_3 + H_2.$$

Most of these reactions have been measured and their rates are thus reasonably well known (Fehsenfeld et al. 1967). In comparing the two reaction schemes, it is interesting to note that H_3O^+ does recombine dissociatively with an electron to form H_2O and OH, but that the analoguous reaction sequence with NH_4^+ seems to take place. The branching ratio of the H_3O^+ dissociative recombination is not accurately known. In this connection, it is also important to note that reactions of H_3^+ with atomic carbon seem to be endothermic (Burt et al. 1970 and references therein), and thus the carbon chemistry in dense molecular clouds can not start this way.

So far we have summarized some basic reactions starting with the cosmic ray ionization of H_2. However cosmic ray ionization of He, which is considerably less abundant in dense clouds than H_2 (about $1/4$) seems to be important for two reasons: firstly, an activation energy barrier (Johnsen and Biondi, 1974) is likely to keep the reaction rates of H^+ with H and H_2 anomalously small (reaction rate $\sim 8 \times 10^{-16}$ cm^3 · sec^{-1}; Sando et al. 1972), and therefore He^+ remains available for the ionization of neutral molecules. Secondly, in most cases, the charge transfer from He^+ to diatomic molecules dissociates them, producing essentially ionized heavy elements, such as C^+, N^+, O^+. The reaction sequence has the general form: (see note added in proof).

$$He + C.R. \rightarrow He^+ + e + C.R.$$

followed by

$$He^+ + AB \rightarrow A^+ + B + He$$

with the specific examples

$$He^+ + CO \rightarrow C^+ + O + He$$
$$He^+ + N_2 \rightarrow N^+ + N + He$$
$$\rightarrow N_2^+ + He$$
$$He^+ + O_2 \rightarrow O^+ + O + He$$
$$He^+ + C_2 \rightarrow C^+ + C + He$$

whereby the latter reaction is of uncertain significance due to the unknown relative abundances of C_2 compared with the CO abundance.

Ionized C^+, N^+, O^+ in turn can be used for the formation of simple ions by reacting with H_2 according to

$$O^+ + H_2 \rightarrow OH^+ + H$$
$$N^+ + H_2 \rightarrow NH^+ + H$$

or they can be involved in formation of more complex molecules.

4.4 Formation of HCN, HNC, CN, and H_2CO

Soon after the discovery of the unidentified molecular emission line U 89.190 and its subsequent identification as HNC is was recognized from the astrophysical data that HNC has comparable abundance to HCN and that this almost equal abundance ratio should be a likely consequence of an ion-molecule reaction mechanism. HCN and HNC are thought to be produced fairly directly from the reaction of C^+ with NH_3 through the following reaction sequences:

$$NH_3 + C^+ \begin{cases} H_2CN^+ + H \\ HCN^+ + H_2 \end{cases}$$

and

$$H_2CN^+ + e \begin{cases} HNC + H \\ CN + H_2 \end{cases}$$

whereby the branching ratio of the latter equation is unknown. However molecular orbital calculations of Conrad and Schaefer 1978 indicate that H_2CN^+ rearranges itself to the energetically lower isometric form $HCNH^+$. Subsequent dissoiative electron recombination forms HNC, HCN in equal abundances since there seems to be no obvious preference in the branching ratio (see note 2 added in proof).

$$H_2CN^+ \rightarrow HCNH^+ + e \begin{cases} HCN + H \\ HNC + H \end{cases}$$

HCN and HNC are likely to be destroyed by reaction with C^+ according to

$$HCN + C^+ \rightarrow C_2N^+ + H$$

followed by

$$C_2N^+ + e \rightarrow C + CN \,.$$

If these reactions are the main destruction process one can estimate a model independent value for the abundance ratios of $([HCN] + [HNC])/[NH_3] \leqq 0.6$ (e.g.

Watson, 1976) for a cloud density of $n_{H_2} \sim 10^6$ cm^{-3}. This ratio can also be determined from the laboratory measured reaction rate (Huntress and Anicich, 1976) to be [HCN]/[NH$_3$] $\sim$ 0.4. Within errors the observationally determined ratios do tend to confirm this value (e.g. Winnewisser et al., (1979)), although it is still not easy to confirm or deny this estimate on the basis of the present interstellar observations.

Formaldehyde, H$_2$CO, is a widely distributed interstellar molecule which occurs in diffuse and in dense molecular clouds. Specific gas phase reactions have been proposed. But formaldehyde has remained a test case molecule where gas-phase reactions cannot completely explain the observed interstellar abundances and where surface reactions might play a role. However, the recently observed HDCO shows that its abundance (Langer et al., (1979)) is not in disagreement with the proposed reaction mechanisms for H$_2$CO. Three mechanisms have been proposed (Watson, (1976), Herbst and Klemperer (1973)):

$$CH_3 + O \rightarrow H_2CO + H$$

which is followed by charge transfer with low ionization metals. Herbst and Klemperer suggested:

$$HCO^+ + H_2 \rightarrow H_3CO^+ + h\nu$$
$$HCO_3^+ + e \rightarrow H_2CO + H \,.$$

The deuterated compound HDCO can be produced by reaction of

$$CH_3^+ + HD \rightarrow CH_2D^+ + H_2 + \Delta E$$

where ΔE is the binding energy difference between the deuterated and the non deuterated species ($\Delta E/k \sim 300\,°K$). This difference in binding energy is generally held responsible for isotope fractionation, also in cases of observed ^{13}C fractionation.

These examples already illuminate the difficulties encountered by gas phase mechanisms even at the level of intermediate sized molecules such as NH$_3$, HCN, HNC or H$_2$CO. With each additional reaction and possible associated branching ratio the uncertainties grow. Despite these difficulties it seems, however, that molecules such as HD, CN, HCO$^+$, HCN, HNC, NH$_3$ and probably also H$_2$CO are preferentially formed through ion-molecule reactions. It seems, however, certain that in the formation of ions and/or radicals the final step in the formation scheme has to occur through gas-phase reaction rather than surface reaction.

5 Formation of Complex Molecules and Carbon Chemistry

5.1 Complex Molecules

The understanding of the formation of complex organic molecules in the interstellar environment is a tantalizing field of interstellar chemistry. Presently it is a very active area for both theoretical and experimental research for it has resisted so far all attempts to even obtain a good qualitative understanding and it promises to remain an area of great challenge and optimism. Despite impressive progress, it has not been possible to overcome the question of whether ion-molecule reactions or surface reactions are the predominant contributors of complex molecules. There are advocates for either mechanism.

The fact that catalytic processes on grain surfaces cannot be neglected does not have its justification in that these processes are easier to understand in the case of complex molecules than for simple ones, but rather that ion-molecule reactions become harder to track, since the uncertainties compound: for converting atoms into complex molecules the number of reaction steps have to increase, whereby reaction rates and branching ratios are less well known. For a large number of supposed reactions rates simply have not even been measured in the laboratory. However progress seems to come in mainly three areas. (i) Extensive studies have been made on reactions of the form

$$AX^+ + BY \rightarrow AB^+ + XY$$

where AB^+ is assumed to be more complex than AX^+.

Applications of this reaction will be specifically discussed for the carbon chain molecules. (ii) the radiative association of larger molecules has been studied by measurements of analogous three body associations (McEwan et al. 1980).

$$A^+ + B + M \rightarrow (AB^+)^* + X \rightarrow AB^+ + M + h\nu$$

with the two distinct intermediate steps

$$(AB^+)^* + M \rightarrow AB^+ + M$$
$$(AB^+)^* \rightarrow AB^+ + h\nu \,.$$

Detailed studies show that the rate coefficients for 3-body association increase with decreasing temperature (Smith and Adams, 1978) probably due to the longer life-time of the excited complex $(AB^+)^*$. Arnold (1979) suggested that radiative association of H_2 to molecular ions may proceed with high rate constants at the low temperatures of dense molecular clouds. (iii) Laboratory observation of a large number of chemical reactions which lead to the formation of complex molecules. Often though, it is not easy to assess the precise relevance of these new data to the interstellar conditions.

On the other hand, the fundamental problem associated with grain production, the return of molecules to the interstellar gas is caused by the low grain temperatures and is certainly not facilitated for complex molecules. On the contrary, the vapour pressure of molecules usually decreases with with increasing molecular complexity as is the case for carbon chain molecules. H_2 overcomes this problem and both CO and N_2 also have sufficient vapour pressure to prevent them from freezing out entirely onto grains. Both molecules serve therefore as reservoirs for atoms and ions through charge transfer reactions discussed earlier.

Infrared observations however have established that local areas of heated dust

(T ~ 40 — 100 K) are wide spread. In addition it has been observationally noted that emission of infrared radiation tends to correlate with molecular line emission from these areas, although notable exceptions exist. In Table 6, the presently known molecules in two selected sources ORIA, and TMC1 are summarized, indicating from the point of view of complexity of molecules, that the temperature does not seem to be the only essential parameter. In the present series, Hayatsu and Anders (1981) review some aspects of complex molecule formation "with an admitted bias towards grain surfaces".

It remains here to point out that various authors (see Hayatsu and Anders 1981 and references therein) have suggested that warm grains could alleviate some of the

Table 6. Molecules Detected in the Orion Molecular cloud and the Taurus Molecular cloud

	Orion (OMC1)	Taurus (TMC1)
Simple hydrides	H_2	
	CH	CH
	OH	OH
	H_2O	NH_3
	H_2S	NNH^+
	NH_3	
	NNH^+	
Oxides, sulfides	CO	CO
	SiO	SO
	O_3	CS
	CS	
	SO	
	SO_2	
	OCS	
	HCS^+	
Acetylene derivatives	CN	CN
	HCN	HCN
	HCCCN	HCCCN
	H_3C-CN	HCCCCCN
	H_2C_2HCN	HCCCCCCCN
	H_3CCH_2CN	HCCCCCCCCCN
	CCH	
	H_3CCCH	CCH
	HNCO	CCCN
	HNC	HNC
	H_3CNH_2	
Aldehydes, Alcohols, Acids and ethers	H_2CO	H_2CO
	H_3CCHO	HCO^+
	H_2CS	
	$HCOOCH_3$	
	CH_3OH	
	$(CH_3)_2O$	
	HCO^+	

The following molecules habe been detected only in the Galactic center sources:
NO, NS, HNO, HCO, CH_3SH, HNCS, HCOOH, CH_2NH, NH_2CHO, CH_3CH_2OH

problems associated with surface mechanisms, in particular the ejection process. Yet even warm grains are far from delivering the complete answer. Observations show that a number of intermediate sized molecules occur in diffuse clouds without known infrared emission (e.g. H_2CO) and secondly large complex chain molecules do occur in cold, dense clouds without any embedded infrared sources and/or small compact H^+ regions. The best studied example of the latter category are probably the Taurus Molecular Clouds in particular TMC1. These sources show a high abundance of complex molecules, expecially the cyanopolyynes, a series of carbon chain molecules with the general formula $HC_{2n+1}N$ with $n = 0, 1, 2, \ldots$

After these more general comments, we would like to discuss within the context of recent laboratory data some of the progress which has been made specifically in the area of complex molecules such as cyanopolyynes. The interstellar carbon chemistry in dense molecular clouds ($n_{H_2} \sim 10^4$ cm^{-3}) is used as an example.

5.2 Carbon Chemistry

Reactions of C^+ with H, H_2 are of fundamental importance to the carbon chemistry in interstellar clouds, and some of the reaction paths prevalent in dense interstellar clouds may also be of significance in the reducing atmospheres of the outer planets. These reactions initiate a complex sequence which produce CH, CH^+ and lead eventually to molecules such as CH, $C_2H_2^+$. These molecules are important precursors to the formation of complex carbon chain molecules found in a large number of astronomical sources.

The explanation of the carbon chemistry took its start with the first attempts by Bates and Spitzer in 1951 to explain the abundances of CH and CH^+. They pointed out that CH^+ in diffuse clouds can be formed by the radiative association process in the gas phase:

$$C^+ + H \rightarrow CH^+ + h\nu \, .$$

The rate coefficient $\alpha \sim 10^{-17}$ cm^3 sec^{-1} has been and is a matter of controversy. Similarly CH can be formed

$$C^+ + H \rightarrow CH + h\nu$$

with a similar rate constant but since carbon is mainly ionized in diffuse clouds, this process does not seem likely. (Reactions for OH, NH have even smaller rate constants). CH can be formed from CH^+ by radiative recombination

$$CH^+ + e \rightarrow CH^*(^2\Sigma^+) \rightarrow CH + h\nu \, .$$

There is also the possibility that the CH^+ ion dissociates by electron capture into C and H:

$$CH^+ + e \rightarrow C + H \qquad (10^{-7} \text{ cm}^3 \text{ sec}^{-1}?) \, .$$

On the other hand photoionization can convert CH into CH^+, i.e.

$$CH + h\nu \rightarrow CH^+(X'\Sigma) + e\,.$$

$$CH + h\nu \rightarrow CH^+(A^3\Pi) + e\,.$$

In dense clouds however reactions with H_2 are of importance: Black and Dalgarno (1973) suggested the radiative association reaction between C^+ and H_2

$$C^+ + H_2 \rightarrow CH_2^+ + h\nu \quad (4 \times 10^{-17}\ cm^3\ sec^{-1})$$

initiating a complex network of reactions leading to larger molecules, as is summarized in Fig. 1. This figure follows the reaction scheme for CH^+ and CH given by Dalgarno 1976 but has been extended to incorporate the starting point of the presumed carbon chain chemistry. Huntress (1977) has published a large number of bimolecular reactions of positive ions which are supposed to be of importance to the chemistry of interstellar clouds, comets and planetary atmospheres of reducing composition.

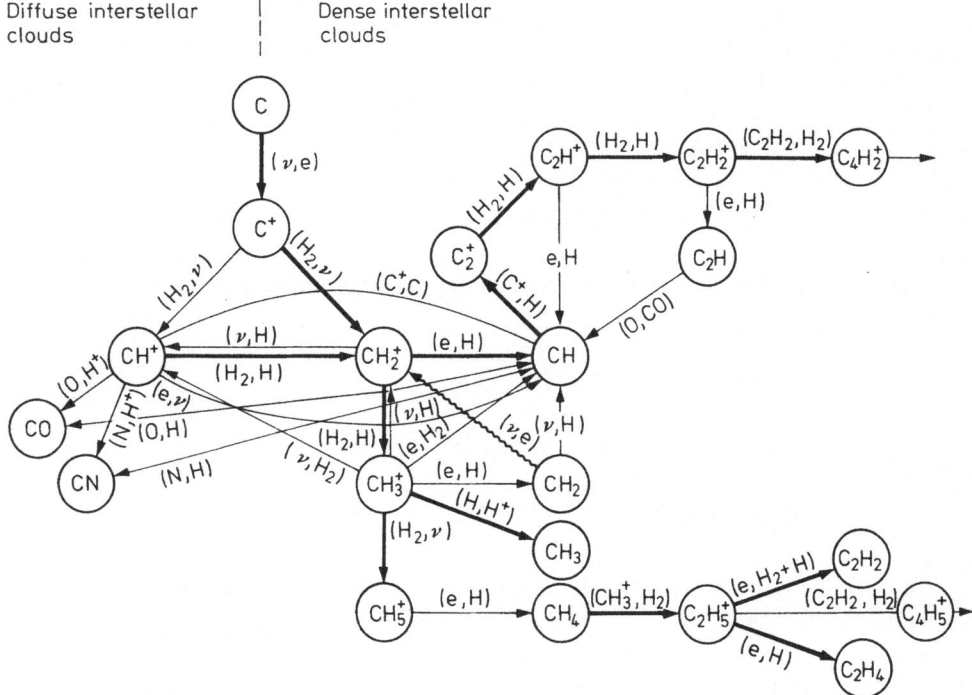

Fig. 1. Interstellar formation scheme illustrating the CH, CH^+, C_nH^+ and higher hydrocarbon cycle. The left side of the reaction cycle pertains to tenous clouds ($n_{N_2} \sim 100\ cm^{-3}$), whereas the right hand side is more appropriate to areas where H_2 is present, i.e. dense molecular clouds ($n_{N_2} \sim 10^4$–$10^6\ cm^{-3}$). The thick arrows indicate assumed preferential reaction paths leading to the higher order hydrocarbons. The following processes are involved: (ν, e) photoionization; (ν, H) photodissociation; (e, ν) radiative recombination; (H); (H_2, ν) radiative association; (e, H), (e, H_2) dissociative electron recombination. (H_2, H) hydrogen abstraction reaction; (C^+, H) charge exchange; (M, M^+) metal charge exchange: metal = Mg, Fe, Ca, Na, . . .

For most of the reactions, rate constants have been determined. In dense clouds, CH^+ is efficiently destroyed by the hydrogen abstraction reaction

$$CH^+ + H_2 \rightarrow CH_2^+ + H \qquad (1 \times 10^{-9} \text{ cm}^3 \text{ sec}^{-1})$$

which is followed by

$$CH_2^+ + e \rightarrow CH + H$$

and

$$CH_2^+ + H_2 \rightarrow CH_3^+ + H \qquad (0.7 \times 10^{-9} \text{ cm}^3 \text{ sec}^{-1})$$

which terminates the hydrogen abstraction sequence, since the reaction $CH_3^+ + H_2$ is endothermic. There are several ways in which CH_3^+ can be destroyed.

Neutralisation of CH_3^+ can be achieved, — as in the case NH_3^+ — by charge exchange reaction with low ionization metals, such as Fe, Ca, Mg, Na, ... In dense clouds there is also the possibility of radiative association followed by dissociative electron recombination to reach neutral CH_4.

The more complex hydrocarbon molecules can be produced either by reaction of CH with C^+ according to the sequence

$$C^+ + CH \rightarrow C_2^+ + H$$
$$C_2^+ + H_2 \rightarrow C_2H^+ + H$$
$$C_2H^+ + H_2 \rightarrow C_2H_2^+ + H$$

where $C_2H_2^+$ can now react by dissociative electron recombination to form C_2H (Watson 1974):

$$C_2H_2^+ + e \rightarrow C_2H + H$$

or in those clouds with an acetylene rich environment:

$$C_2H_2^+ + C_2H_2 \rightarrow C_4H_2^+ + H_2 \, .$$

Another likely possibility for the formation of hydrocarbon molecules starts with methane CH_4 and CH_3^+

$$CH_4 + CH_3^+ \rightarrow C_2H_5^+ + H_2$$

which then by dissociative electron recombination could produce the simplest hydrocarbon chain molecules:

$$C_2H_5^+ + e \nearrow^{C_2H_2 + H_2 + H}_{\searrow C_2H_4 + H}$$

Follow-up steps in this series would be further reactions between CH_3^+ and C_2H_2, C_2H_4 and C_2H_6. These initial steps leading to the formation of complex hydrocarbon molecules are indicated in Fig. 1.

It remains however to mention that the reactive CH_3^+ ion can react with other atoms and molecules. Some of these reactions are with N, NH_3 or H_2CO leading essentially to the formation of HCN and HCO^+ (Huntress 1977):

$$CH_3^+ + N \nearrow^{HCN^+ + H_2}_{\searrow H_2CN^+ + H}$$

$$CH_3^+ + H_2CO \rightarrow HCO^+ + CH_4$$

$$CH_3^+ + NH_3 \rightarrow CH_2NH_2^+ + H_2$$

$$CH_2NH_2^+ \nearrow^{HCN + H_2 + H}_{\searrow CN + 2H_2}$$

This presents a short summary of the reactions which are supposed to be important gas-phase reactions leading to the complex hydrocarbon molecules, and some of their derivatives. In interstellar space the cyanopolyynes constitute an important class, since only molecules with a permanent dipole moment can be observed in radioastronomical measurements. Thus hydrocarbons which have no dipole moment elude radioastronomical detection.

The detection of cyanopolyyne molecules in interstellar space and their wide distribution seem to have revived the discussion of the fundamental question of how and where molecules form. The discussion centers around the question whether the origin of the molecules is to be interpreted as an *in situ* synthesis from interstellar gas or if the opposite hypothesis is correct. In this picture the observed large organic molecules are dissociation products of much more complex molecules or, in effect, they are the result of spallation of interstellar grains which are composed of or covered by organic matter. In the present volume on cosmochemistry, Hayatsu and Anders (1981) have discussed organic compounds in meteorites and their origin, and have given arguments for the formation of cyanopolyynes on grain surfaces.

Before discussing the possibility of gas-phase cyanopolyyne chemistry, it seems necessary to summarize the present status of carbon chain molecule detections in interstellar space. Table 7 presents an overview of where these molecules are found and their respective abundances. These tables are an abbreviated update from Table 1 taken from Winnewisser and Walmsley (1979). It is seen that these molecules are found essentially in every type of molecular cloud from the cold dark cloud to the warm circumstellar environment, underlining the trend which has been observed over the past few years: namely that complex organic molecules are not limited to a few sources only (in particular to the galactic center sources) but that they are spread over sources with rather different physical conditions. A qualifying statement may be in order here.

In contrast to the dark cloud chemistry, the molecules in circumstellar envelopes (IRC +10216) seem to be created continuously in a small, high temperature high density layer- which allow fast thermodynamic equilibrium- and subsequently expelled into the lower density cool envelope. There they are observed with an

Table 7. Distribution of carbon chain molecules

Molecule	Cloud type				clouds with assoc. H⁺ region			Circum stellar	
	Dark clouds								
	TMC1	L 1544	L 183	ϱ Oph.	Sgr B2	Ori A	W 51	IRC + 10216	CRL 2688
CN	•				•	•	•	•	
HCN	•		•	•	•	•	•	•	•
HC$_3$N	•	•	•	•	•	•	•	•	
HC$_5$N	•	•	•	•	•	•		•	
HC$_7$N	•				•			•	
HC$_9$N	•				•			•	
C$_2$H	•				•	•		•	
C$_4$H	•							•	
C$_3$N	•							•	
HCCH								•	
C$_2$								•	•
C$_3$								•	•

Molecular abundances of carbon chain molecules in selected sources (Winnewisser a. Walmsley 1979)

Source	Distance d (kpc)	Density n(H$_2$) (cm^{-3})	Temperature T$_K$ (K)	Molecular column densities (cm^{-2})									
				CN	HCN	HC$_3$N	HC$_5$N	HC$_7$N	HC$_9$N	C$_2$H	C$_4$H	C$_3$N	C$_2$H$_2$
TMC1	0.1	3×10^4	10	$\sim 10^{13}$		6×10^{13}	7×10^{13}	2×10^{13}	0.3×10^{13}				
L 183	0.1	3×10^4	10	$<3.6 \times 10^{13}$	$\sim 3 \times 10^{12}$	$\sim 10^{12}$				$<6 \times 10^{12}$		7×10^{12}	
ORI A	0.45	$\sim 10^5$	50–70	3×10^{13} -9×10^{14}	10^{15}	2×10^{13}				$\left\{\begin{array}{l}2 \times 10^{14}\\3 \times 10^{15}\end{array}\right.$			
SGR B2	10	$\sim 10^4\text{--}10^6$	50	3×10^{14}		2×10^{14}	2×10^{14}						
IRC + 10216	0.29	~ 300		1×10^{15}	10^{15}	2×10^{14}	4×10^{14}	10^{14}		$\sim 5 \times 10^{14}$	$\left\{\begin{array}{l}4 \times 10^{14}\\3 \times 10^{15}\end{array}\right.$	$\left\{\begin{array}{l}1 \times 10^{14}\\8 \times 10^{14}\end{array}\right.$	3×10^{19}

essentially "frozen-in" chemical equilibrium. It seems that grains are produced and ejected simultaneously with the molecules. In this picture, the large organic molecules might be spallation products of interstellar grains.

However, the present discussion pertains to dark cloud chemistry. The experimental interstellar observations clearly indicate that the distribution of carbon chain molecules is correlated, and that the column densities of the longer chain members decreases about linearly with increasing chain length. Several mechanisms have been proposed for the chain building. For cool dark clouds Churchwell et al. (1978) and in further detail Walmsley et al. (1979) have proposed a formation scheme by which the longer chain molecules are formed via the acetylene "backbone" reaction:

$$C_2H_x^+ + C_2H_2 \rightarrow C_4H_y^+ + C_2H_2 \rightarrow C_6H_z^+ + C_2H_2 \rightarrow \ldots$$

where then the appropriate cyanopolyyne species is formed from $C_2H_2^+$, $C_4H_4^+$, ... with notably CN or HCN as abundant nitrogen bearing molecules:

$$C_2H_2^+ + HCN \nearrow^{HC_3NH^+ + H}_{\searrow H_2C_3N^+ + H_2}$$

followed by dissociative electron recombination. Huntress et al. 1980 discusses in further detail the synthesis of very complex organic molecules by ion-molecule reactions.

Huntress 1977 also points out that the reaction is slow and could have a significant temperature dependence. However if this mechanism contributes the major part of the cyanopolyyne chemistry than one has to conclude (Winnewisser et al. 1980) (i) that the abundance of the cyanopolyyne molecules will decrease with increasing length of the carbon chain, (ii) that long chain molecules with other functional groups such as CH_3, NH_2, ... should be observable (iii) that molecules with no permanent dipole moment such as the hydrocarbons (saturated and unsaturated) acetylene, HCCH, diacetylene, HCCCCH, ... should have high abundance in the interstellar medium (iv) that unstable species such as HCCN, H_2CCN should also be abundant in the appropriate molecular clouds. Some of the unidentified interstellar lines could have these reactive species as their carrier. Similar to interstellar HCN higher cyanopolyyne members are likely to be destroyed by C^+.

It was noticed during the acetylene-hydrogen cyanide discharge experiments that they are always accompanied by a fairly rapid formation of a brown deposit. A similar polymerization product has been observed in the pure acetylene discharge of Vasile and Smolinsky. Sagan and Khare (1979) have analysed a presumably similar residue, the "intractable polymer", or what they call "tholin" (Greek "muddy"), which they obtained by discharging CH_4, C_2H_2, NH_3, H_2O and H_2CO. They find that their "tholins" contain a large fraction of the presently known interstellar molecules and argue that tholins are a major constituent of the interstellar grains. The observed large interstellar molecules could be produced by spallation from such grains. Molecular abundance arguments are used to favour this view rather than in situ synthesis from interstellar gas. However, the largest organic molecules, the cyanopolyynes, show a decline in abundance with increasing length (Broten et al., 1978). In this connection it is highly interesting to note that Sakata et al. (1977) have found spectroscopic evidence that "an extract of organic material" (mainly aromatic polymers from the Murchison carbonaceous chondrite) shows absorption features near 2000 Å with a half-width ~ 300 Å. Organic molecules with conjugated multiple bonds such as $C \equiv C - C \equiv N$, $H - C \equiv C - C$, $C = C = O$ are known to absorb near 2200 A. Douglas (1977) has suggested that the diffuse interstellar bands observed by optical astronomy could be caused by long carbon chain molecules C_n ($5 \leqq n \leqq 15$) which would not be easily photodissociated, since they can transform excess energy by internal conversion.

Although there seems to be no doubt that such molecules are abundant in the brown deposit of the discharge products or the "tholins", or more colloquially the "laboratory grains", the coincidence between their absorption and the observed interstellar spectra of sources such as NGC7538E, NGC7027 are at best suggestive, but not conclusive. Thus the idea that grains consist of tholins or a variety of organic compounds, polysaccharides or carbonaceous compounds is speculation. In fact, quite to the contrary, Duley and Williams (1979) very recently concluded that there is little spectroscopic evidence to support the contention that much of the interstellar dust consists of organic material. In particular, grains made of organic material might be expected to show the C-H stretching vibration between 3.3 and 3.4 μm. The carbynes, a triply bonded, linear allotrope of elemental carbon could be an interesting alternative grain (see for example Hayatsu and Anders, this issue).

Although Fischer-Tropsch-type reactions have so far failed to produce the heavier cyanopolyynes — the reason may be technical (Hayatsu and Anders this issue) — HC_3N and HC_5N have been produced in a discharge, starting from HCCH and HCN (Winnewisser et al. 1978). More recently it was noticed that in other (presumed) gas-phase reactions HC_3N can be observed as well: decomposition of vinylcyanide by action of a discharge splits off hydrogen to yield the unsaturated HC_3N (Winnewisser et al. 1981):

$$\begin{array}{c} H \qquad H \\ \diagdown \quad \diagup \\ CC \qquad \rightarrow HCC - CN + H_2 \\ \diagup \quad \diagdown \\ H \qquad CN \end{array}$$

A similar reaction of vinylcyanide H_2C_2HCN with HCCH yields HC_5N. In this context the reaction of acetylene, HCCH, with fulminic acid, HCNO, is of interest, since it produces HCCCN but not HCC-CNO (Winnewisser et al. 1981):

$$HCCH + HCNO \xrightarrow{\text{r.f. discharge}} HCCCN + H_2O$$
$$\xrightarrow{} HCC - CNO + 2H$$

In a reducing environment hydrogen atoms will scavenge the oxygen atoms from molecules with NO bonds. This might be the reason that molecules, such as HCNO, N_2O are absent in interstellar space or have fairly low abundances such as NO.

Hayatsu and Anders (this volume) point out from a comparison of the relative abundance pattern of cyanopolyynes $H(C=C)_nN$ ($n = 1, 2, 3, 4$) with that of alcohols $C_nH_{2n-1}OH$ (Fig. 13 of their contribution) that the probability of the cyano polyyne carbon chains growing by one C unit is equal to the "Fischer-Tropsch-type" alcohols. Clearly if the proposed gas-phase mechanism operative in interstellar clouds is the only contributor to the formation of the cyanopolyyne chemistry, and if this mechanism feeds only on itself without other contributing reactions, one would expect an exponential decrease in the abundance. This is not observed. However, several searches for $HC_{11}N$ has so far remained negative. The recent discovery of the high abundance of DC_3N in TMC1 suggests that gas phase reactions must be involved.

In this sense one can only agree with the conclusion of Hayatsu and Anders "that a truly comprehensive review-study of *all* relevant reactions" could reveal the relative importance of gas phase versus surface reactions. It may then turn out that the "truth" is achieved by employing both mechanisms at different stages during the interstellar formation process.

In summary it seems that in situ synthesis of long chain carbon molecules is presently the most convincing of the various formation mechanisms. In particular, spallation of organic grains seems rather unlikely in the cold dark clouds such as TMC 1. We note, incidentally, that the dark clouds produce an absolutely "clean chemistry", in the sense that many types of reactions which occur in terrestrial chemistry are excluded. Shocks, for example, appear not to be present if one can judge from the observed narrow line profiles. The gas is very quiescent and cold. On the other hand, ions such as HCO^+ and N_2H^+ (Guélin et al., 1977) are present within these condensations and we therefore think that molecular-ion production schemes should be investigated further. Some consequences of an ion-molecule formation scheme for the cyanopolyynes have already been discussed. Green and Herbst (1979) point out that even if molecules are produced, say, on grain surfaces, secondary processing by molecular ions is liable to have observable consequences. For example, if, as we have suggested, acetylene, HCCH, is very abundant in TMC 1, then it is likely that its isomer H_2C—C (vinylidene) will be produced via the sequence

$$C_2H_2 + H_3^+ \rightarrow C_2H_3^+ + H_2$$

$$C_2H_3^+ + e \begin{cases} \nearrow HC \equiv CH + H \\ \searrow H_2C = C + H \end{cases}$$

Vinylidene, unlike acetylene, should have a dipole moment and hence may be observable at radio wavelengths. In a similar fashion, one might expect that a high abundance of HC_3N would go together with a large concentration of the isomers HC_2NC and C_3NH. An obvious goal of future research is to attempt to detect isomers and determine their relationship to those already known. The isomers of isocyanic acid, HNCO, cyanuric acid, HOCN, and fulminic acid, HCNO, furnish a particularly interesting example: HNCO has been detected in interstellar space, a tentative assignment of interstellar HOCN exists (Guélin et al., 1981), and HCNO has not been found in agreement with the generally low abundance of molecules with one or more NO bonds.

6 Molecules in Comets and Planetary Atmospheres

Although it is clearly beyond the scope of this review to discuss in detail the composition and chemistry of comets and planetary atmospheres, it seems proper to give a short summary of the molecular species which have been detected there. From Table 8 and 9, it is evident that remarkable similarities exist between the composition of the interstellar medium, comets and the atmospheres of the outer planets which is, in part, based on the reducing environment (with the exception of the inner planets) and the carbon based chemistry. The ability of carbon atoms to combine with as many as four other atoms and its ability to form long chains in which the carbon atoms form the "backbone" of the entire molecule gives them their fundamental importance. The list of detected molecules (interstellar, cometary and planetary) contains a large number of molecules with one or more carbon atoms. In fact, the most complex molecules safely identified in interstellar space are carbon based molecules and even in the highly reducing atmospheres of Jupiter and Saturn hydrocarbons have been identified. Thus the carbon chemistry in interstellar space is not only interesting in its own right but also because the more complex interstellar molecules are found in meteorite materials, in comets, and in the primordial atmospheres of the outer planets. They also resemble closely the basic building blocks of living matter on earth. Despite these similarities, it is not clear, however, how the existence of interstellar molecules relates to the existence of molecules in comets and planetary atmospheres. Recently, Larson 1980 and Prinn and Owen 1976 (and references therein) have summarized in detail the atmospheres of the outer planets. From Table 9, it becomes clear that next to molecular hydrogen, methane CH_4, is the major constituent in the atmospheres of the outer planets, and its photodissociation yields CH and CH_2 radicals in a ratio of about 8 to 92. CH_2 reacts with H_2 to form CH_3 which, in turn, captures H to form CH_4.

Table 8. Observed composition of comets (Whipple and Huebner (1976))

Head: H, C, C_2, C_3, CH, CN, $^{12}C^{13}C$, HCN, CH_3CN, NH, NH_2, O, OH, H_2O, Na, K, Ca, V, Cr, Mn, Fe, CO, Ni, Cu.

Tail: CH^+, CO^+, CO_2^+, N_2^+, OH^+, H_2O^+, Ca^+.

Continuum from particles including Silicate 10- and 18 μm bands in head and tail.

Table 9. Molecules detected in Planetary Atmospheres

Planet	Molecules
Mercury	No definite identification
Venus	CO_2 (96%), N_2 (3.5%), H_2O, HCl, HF, H_2SO (droplets), Ar (90 Earth Atm.)
Earth	N_2 (80%), O_2, $_{=2}O$, CO_2, CH_4, H_2, CO; Ar, Xe, Ne, Kr, N_2O, O, O_2, O_3
Mars	CO_2 (95%), N_2 (2.7%), Ar, O_2, H_2O (0.01 Earth Atm.)
Jupiter	H_2, He, CH_4, NH_3, H_2O, PH_3, GeH_4, CO, C_2H_2, C_2H_6
Saturn	H_2, CH_4, NH_3, PH_3, C_2H_2, C_2H_6; (He assumed)
Uranus	H_2, CH_4
Neptun	H_2, CH_4
Pluto	No identifications

About 20% of the photo-dissociated CH_4 is then converted to the hydrocarbon chain molecules C_2H_2, C_2H_4 and C_2H_6, two of which have been detected in the jovian atmosphere. In the ionosphere of Jupiter and in the regions of discharges in the cloudy zones of the atmosphere, the higher-order hydrocarbon ions are the likely precursors of C_2H_2, C_2H_4, ... which could form according to the reactions:

$$C_2H_5^+ + C_2H_4 \rightarrow (C_4H_9^+)^* \rightarrow C_4H_5^+ + CH_4$$
$$C_2H_5^+ + C_2H_2 \rightarrow (C_4H_7^+)^* \rightarrow C_4H_5^+ + H_2$$

followed by dissociative electron recombination which is similar to the interstellar mechanism. The hydrocarbon reaction scheme for Jupiter, illustrated in Fig. 2 (according to Prim and Owen), reveals the similarities between the interstellar and planetary chemistry in a reducing environment. The longer hydrocarbons produced in the ionospheres and higher atmospheres of the giant planets Jupiter and Saturn are convectionally transported down to the hot lower atmosphere. There they are converted back into CH_4 most likely by thermochemical reactions. In this sense it is "raining crude oil" in the atmospheres of Jupiter and Saturn. The hydrogen abundance in Uranus and Neptune is lower than that of Jupiter and Saturn, indicating that they have lost some hydrogen and helium in the course of their

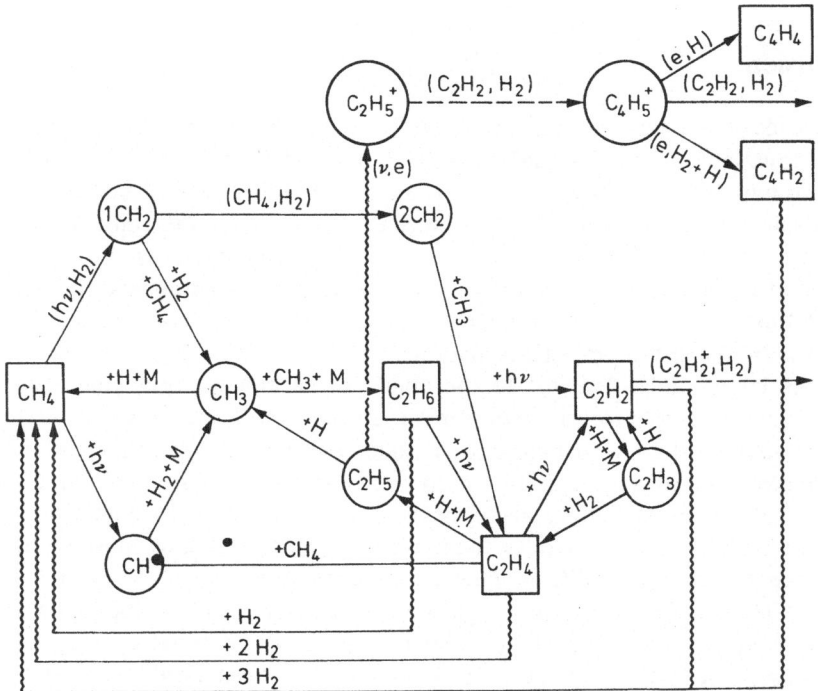

Fig. 2. Planetary methane and higher order hydrocarbon cycle in Jupiter's mesosphere and stratosphere (from Prinn and Owen, (1976)) and upper atmosphere. Vertical transport is represented by vertical wavy lines. Squares enclose stable molecules, whereas ions, radicals and unstable molecules are encircled.

history. Contrary to the reducing primordial atmospheres of the outer planets, the inner planets (notably Venus and Earth) have oxidizing secondary atmospheres, which were made after the planets had been formed and most of primordial hydrogen and helium was lost.

Comets offer the chance, each time they come to close encounter with the sun, to investigate early primitive solar system material which presumably was formed directly from the gas and dust of the solar nebula. It is assumed that the nucleus of comets consists of large grains of rocky material (meteoroidal material) which is mixed and/or covered with ices, molecules and atoms which are bonded to the ices (clathrates). Although the true nucleus of a comet is rarely observed, the comets are extensively being studied by the material they lose under the action of solar radiation during close solar encounter, i.e. at distances of less than 5 AU (1 AU = 1 astronomical unit = 1.5×10^{13} cm). The matter comets lose in this process are mainly gases and meteorites. Table 8 summarizes all so far detected cometary molecules. It is seen that the head of the comet contains molecules, radicals and atoms, whreas the tail shows only ions. Practically all of these molecules and ions are also interstellar species. For further detail on cometary spectra and the prevailing physical processes, the reader is referred to an excellent review by Whipple and Huebner (1976). All of the present knowledge has come from the analysis of cometary spectra but the planned space probe mission (if carried out) to Halley's comet in 1986 would be of superb scientific value.

7 Conclusion

The past decade of interstellar research has established that the most widely distributed and most abundant interstellar molecules (OH, CH, H_2O, H_2CO, HCN, NH_3) are found in molecular clouds within our Galaxy and in other galactic systems, i.e. in extragalactic sources such as the Andromeda Nebula (M31) the spiral Galaxy NGC 253 and others. On the galactic scale the interstellar molecules are found in a large variety of astronomical objects with widely different physical parameters, ranging from the diffuse clouds and the cold dark clouds to the dense hot circumstellar environment. In the list of molecules detected to date, carbon based molecules dominate the larger molecules, due to the ability of carbon to form long chain molecules. Oxygen and nitrogen with similar cosmic abundance do not share this ability; at most they form hydrogenperoxide, HOOH, or hydrazin, HNNH, as chain molecules. Silicon, with lower cosmic abundance than C, N and O does form silicon chains (mainly a silicon-oxygen bond). However, its high affinity to oxygen rather forms silicon oxides than silicon hydrites and thus most of the silicon will be locked up as SiO in interstellar space. Even in a very reducing atmosphere SiH_4 is only formed as a high temperature species (T $\sim$ 1000 K) and one does not therefore expect to find SiH in interstellar space in detectable quantities.

Carbon based chemistry (organic chemistry) has thus been established to be of fundamental importance in interstellar molecular clouds. Similarly the observed composition of comets is dominated by carbon bearing molecules, and in the reducing atmospheres of Jupiter and Saturn the carbon chain molecules C_2H_2, C_2H_6 have been detected.

Interstellar molecules are detected at the position where they are formed. Their formation mechanism is usually modelled for a steady-state situation, although their abundances are not in thermodynamic equilibrium. Cosmic rays and ultraviolet radiation prevent equilibrium from being reached. Cosmic ray ionization is seen as the driving force for a large number of chemical reactions.

Catalytic processes and gas-phase-reactions are required to explain molecule formation. The rate at which atoms stick to the surface of interstellar dust grains seems to be comparable to the reaction rates of various important gas phase reactions. Although there is general agreement that H_2 is formed on grain surfaces, HD is the strongest evidence for gas phase reaction mechanisms. In fact, gas-phase models have had the better of the arguments so far, at least for the explanation of the abundance of simple molecules, notably the ions HCO^+, N_2H^+ and HCN, HNC, CCH. For this simple class of gas phase reactions, the ion-molecule model produces predictable results. For ion-molecule reactions to be fast they must be exothermic and have no activation energy barrier. Already for intermediate sized molecules such as NH_3, H_2CO predictions based on the ion-molecule scheme become uncertain. The formation of complex molecules is presently a great challenge in interstellar chemistry. Although both mechanisms, grain-surface and ion-molecule reactions are thought to contribute, their main problems remain: uncertainties concerning the surface processes are mainly connected with the ejection mechanism on one hand, branching ratios and reaction rates on the other require considerably more detailed, but extremely interesting and rewarding laboratory investigations. In addition, the many unidentified lines detected to date in interstellar space are likely to be caused by short lived reaction intermediates or by complex molecules not yet synthesized in the laboratory. For their proper identification precise laboratory microwave frequencies are required. These studies have now being extended to high resolution infrared studies of molecules.

8 References

Aanestad, Per A., Purcell, E. M.: Ann. Rev. of Astron. Astrophys. *11*, 309 (1973).

Adams, N. G., Smith, D.: IAU No. 87, Interstellar Molecules (B. H. Andrew, ed.) p. 311 (1980).

Arnold, F.: Proc. XXI Liège Colloquium, 494 (1977).

Bates, D. R., Spitzer, L.: Astrophys. J. *113*, 441 (1951).

Black, J. H., Dalgarno, A.: Astrophys. (Letters), *15*, L79 (1973).

Black, J. H., Dalgarno, A.: Astrophys. J. Suppl. *34*, 405 (1977).

Black, J. H., Hartquist, T. W., Dalgarno, A.: Astrophys. J. *224*, 448 (1979).

Broton, N. W., Oka, T., Avery, L. W., MacLeod, J. M., Kroto, H. W.: Astrophys. J. *223*, L105 (1978).

Burt, J. A., Dunn, J. L., McEwan, J. M., Sutton, M. M., Roche, A. E., Schiff, H. I.: J. Chem. Phys. *52*, 6062 (1970).

Conrad, M. P., Schaefer, H. F.: Nature *274*, 456 (1978).

Dalgarno, A.: Atomic Processes and Applications (ed. P. G. Burke, B. L. Moiseiwitsch), p. 109, Amsterdam: North Holland (1976).

Dalgarno, A.: IAU No. 87, Interstellar Molecules (B. H. Andrew, ed.) 273 (1980).

Dalgarno, A., Black, J. H.: Rep. Prog. Phys. *39*, 573 (1976).

Dalgarno, A., McCray, R. A.: Astrophys. J. *181*, 95 (1973).

Douglas, A. E.: Nature *269*, 130 (1977).

Duley, W. W., Williams, D. A.: Nature *277*, 40 (1979).

Elitzur, M., de Jong, T.: Astrophys. J. (Letters) *222*, L141 (1978).

Fehsenfeld, F. C., Schmeltekopf, A. L., Dunkin, D. B., Ferguson, E. E.: ESSA tech. Rep. ERL 135-AL3 (U.S. Dep. Commerce) (1969).

Fehsenfeld, F. C., Dunkin, D. B., Ferguson, E. E., Albritton, D. L.: Astrophys. J. *183*, L25 (1973).

Green, S., Herbst, E.: Astrophys. J. *229*, 121 (1979).

Guélin, M., Langer, W. D., Snell, R. L., Wootten, H. A.: Astrophys. J. *217*, L165 (1977).

Hartquist, T. W., Black, J. H., Dalgarno, A.: Mon. Not. Roy. Astr. Soc. *185*, 643 (1978).

Hartquist, T. W., Oppenheimer, M., Dalgarno, A.: Astrophys. J. (in press) (1981).

Herbst, E., Klemperer, W.: Astrophys. J. *185*, 505 (1973).

Herbst, E., Klemperer, W.: Physics Today *29*, 32 (1976).

Herbst, E.: Astrophys. J. *222*, 508 (1978).

Herbst, E.: in Protostars and Planets, p. 88 (T. Gehrels ed.), Univ. of Arizona Press, Tucson (1978).

Hollenbach, D. J., Salpeter, E. E.: Astrophys. J. *163*, 155 (1971).

Hollenbach, D. J., Werner, M. W., Salpeter, E. E.: Astrophys. J. *163*, 165 (1971).

Hollenbach, D. J., McKee, C. F.: Astrophys. J. Suppl. (in press) (1980).

Hunter, D. A., Watson, W. D.: Astrophys. J. *226*, 447 (1978).

Huntress, W. T., Jr.: Astrophys. J. Suppl. *33*, 495 (1976).

Huntress, W. T., Jr., Prasad, S. S., Mitchell, G. F.: IAU No. 87, Interstellar Molecules (B. H. Andrew, ed.) p. 331 (1980).

Iglesias, E. R., Silk, J.: Astrophys. J. *226*, 851 (1978).

Johnsen, R., Biondi, M. A.: J. Chem. Phys. *61*, 2112 (1974).

Jura, A.: Astrophys. J. *197*, 581 and references therein (1975).

Langer, W. D., Frerking, M. A., Linke, R. A., Wilson, R. W.: Astrophys. J. *232*, L65 (1979).

McCabe, E. M., Smith, R. C., Clegg, R. E. S.: IAU Interstellar Molecules p. 497–502, (B. H. Andrew, ed.) (1980).

McDaniel, E. W., Cermak, V., Dalgarno, A., Ferguson, E. E., Friedman, L.: Ion-Molecule Reactions, Wiley-Interscience, New York (1970).

McEwan, M. J., Anicich, V. G., Huntress, W. T., Jr., Kemperer, P. R., Bowers, M. T.: in IAU No. 87, Interstellar Molecules, (B. H. Andrew, ed.), p. 305 (1980).

Prinn, R. G., Owen, T.: in Jupiter (T. Gehrels, ed.) p. 319 (1976).

Sagan, C., Khare, B. N.: Nature *277*, 102 (1979).

Sakata, A., Nakagawa, N., Iguchi, T., Isobe, S., Morimoto, M., Hoyle, F., Wickramasinghe, N. C.: Nature *266*, 241 (1977).

Sando, K. M., Cohen, R., Dalgarno, A.: Conf. Electronic and Atomic Collisions (T. R. Govers, F. J. de Heer eds.) North Holland, Amsterdam, p. 973 (1972).

Savage, B. D., Mathis, J. S.: Ann. Rev. Astron. Astrophys. *17*, 73–111 (1979).

Schmeltekopf, A. L., Fehsenfeld, F. C., Ferguson, E. E.: Astrophys. J. *148*, L155 (1967).

Smith, D., Adams, N. G.: Kinetics of Ion Molecule Reactions (P. Auloos, ed.), Plenum Press, New York, p. 345 (1978).

Snow, T. P., Jr.: IAU No. 87, Interstellar Molecules p. 247 (B. H. Andrew, ed.) (1980).

Thaddeus, P., Guélin, M., Linke, R. A.: Astrophys. J. (in press) (1981).

Vasile, M. J., Smolinsky, G.: Int. J. Mass Spect. Ion Phys. *24*, 11 (1977).

Watson, W. D.: Astrophys. J. *182*, L69 (1973).

Watson, W. D.: Astrophys. J. *183*, L17 (1973).

Watson, W. D.: Astrophys. J. *188*, 35 (1974).

Watson, W. D.: Rev. Mod. Phys. *48*, 513 (1976).

Watson, W. D.: Ann. Rev. Astron. Astrophys. *16*, 585 (1978).

Watson, W. D.: Proc. Liège Int. Astrophys. Symp. 21st, p. 526 (1978).

Watson, W. D., Christiensen, R. B., Deissler, R. J.: Astron. a. Astrophys. *69*, 159 (1978).

Watson, W. D.: in IAU No. 87, Interstellar Molecules (B. H. Andrew ed.) p. 341, (1980).

Whipple, F. L., Huebner, W. F.: Ann. Rev. Astron. Astrophys. *14*, 143 (1976).

Winnewisser, G., Churchwell, E., Walmsley, C. M.: in Modern Aspects of Microwave Spectroscopy, (G. W. Chantry ed.), 313–503 Academic Press, London (1979).

Winnewisser, G., Mezger, P. G., Breuer, H. D.: Topics in Current Chemistry, Vol. 44, Springer-Verlag, Berlin, Heidelberg, New York (1974).

Winnewisser, G., Walmsley, C. M.: Astrophys. a. Space Sci. *65*, 83 (1979)

Winnewisser, G., Toelle, F., Ungerechts, H., and Walmsley, C. M.: in IAU No. 87, in Inter-
   stellar Molecules (B. H. Andrew, ed.), p. 59 (1980).
Winnewisser, G., Winnewisser, M., Christiansen, J. J.: Astron. a. Astrophys. (in press) (1981).
York, D. G., Rogerson, J. B. Jr.: Astrophys. J. *203*, 378 (1976).
Zuckerman, B.: Ann. Rev. Astron.-Astrophys. *18*, 263 (1980).

Received April 9, 1981

Note 1 Added in Proof

Johnsen et al. (1980) have measured the rate coefficient for $He^+ + H_2$ to be
10^{-13} cm^3 sec^{-1} down to 78 K. Thus this reaction cannot be neglected any longer
compared to the ionization of C, N, O ... by He^+.

Johnsen, R., Chen, A., Biondi, M. A.: J. Chem. Phys. *72*, 3085 (1980).

Note 2 Added in Proof

The reaction of C^+ with NH_3 might lead preferentially to HNC via the isomer
H_2NC^+ (Allen et al. (1980), Brown, (1977)) which lies ~ 2 eV above the linear
isomer $HNCH^+$. The stable triplet configuration H_2NC^+ might radiatively decay
into singlet H_2NC^+ rather than isomerize to the linear form.

In this connection it worthwhile to note that HCN and HNC can be converted
into each other by reactions of the form

$$\left.\begin{array}{l} HCO^+ \\ N_2H^+ \\ H_3^+ \end{array}\right\} + HCN, HNC \quad HNCH^+ + \left\{\begin{array}{l} CO \\ N_2 \\ H_2 \end{array}\right.$$

or by proton exchange

$$HNC + H^+ \rightarrow HCN + H^+$$

Local disturbances in the clouds might favour one or the other molecule. In the
Orion ridge Goldsmith et al. (1981) find an overabundance of HCN which might be
an indication of shock influence.

Allen, M., Goddard, J. D., Schaefer, H. F., III., J. Chem. Phys. *73*, 3266, (1980).
Brown, R. D., Nature *270*, 39 (1977).
Goldsmith, P. F., Langer, W. D., Ellder, J., Irvine, W., Kollberg, E., (1981), preprint.

Chemistry in Comets

Rhea Lüst

Max-Planck-Institut für Physik und Astrophysik Institut für Astrophysik
Karl-Schwarzschild-Str. 1, D-8046 Garching, FRG

Table of Contents

1 Introduction

The chemical composition of comets and the processes taking place in their comas and tails during their active phase near the Sun are very complex and not as yet known in their details though investigations from sounding rockets and satellites in the ultraviolet spectral regions and measurements with radio telescopes have provided a wealth of new results during the last two decades. For a better understanding of the problems and advances, an introductory chapter shall give a brief review of the history and more recent evolution of cometary research.

Observations of comets have been reported as early as several centuries before Christ, mainly by Chinese historians. Up to the 15th and 16th centuries, their true nature was unknown, and they were regarded as transient phenomena of the "sublunar sphere", as the terrestrial atmosphere and the close neighborhood of the Earth were called. Their apparitions, unexpected and sometimes very spectacular, were accompanied by superstitious fear, and they were regarded as heralds or even the cause of evil. And because our history is full of wars, epidemics and natural disasters, coincidences with comet apparitions have always been easy to find. Even after their cosmic position had become clearer, superstition did not really come to an end, and we can find it up to the present time. During the last apparition of Comet Halley in 1910, people were warned in various newspapers to beware of poisonous gases since the Earth passed through the outer parts of the tail which contains indeed carbon monoxide, though in such an extremely low concentration that the "comet pills" sold by clever merchants against the noxious vapors were certainly not necessary.

The first scientific observations of comets are linked with the names of Regiomontan, Apiano and especially Tycho Brahe. It was the latter who found from precise observations of a bright comet visible at the end of 1577 that this object was further away from the Earth than the Moon, and he concluded that comets belong to the "translunar sphere", pervading the region between the planets. About 100 years later, Halley calculated the first cometary orbits by applying Kepler's laws of planetary motions, and he succeeded in identifying the bright comet of 1682 — later becoming so famous as "Comet Halley" — with two former comets of 1531 and 1607. He could not witness the next apparition of this comet in 1759, but it turned out that he had correctly predicted the date of its return, and he can thus be regarded as the "father" of the periodic comets. From that time, the comets can be grouped into those which appear only once on very elongated or nearly parabolic orbits and the periodic comets which return after a certain number of years.

About 100 years ago, the means of photography and spectroscopy opened a new aera in astronomy and it became possible to try a first crude chemical analysis of the comets. It was found that their spectra consist of a narrow continuum which was correctly interpreted as a reflection of the sunlight from the solid nucleus and from small dust particles in the coma, and of a line and band spectrum emitted by the gases of the cometary atmosphere. Already in 1911, K. Schwarzschild and E. Kron, using observations of Comet Halley, explained these emissions by resonance fluorescence of the solar radiation from various molecules. Identifications yielded the presence of the elements H, C, O, and N in various compounds. The strongest bands were the Swan system of C_2 in the visual and the CN emissions in the blue and violet spectral regions. It is this radical which determines the extension of the coma in the photographic region, as the somewhat smaller visual coma is determined by the Swan bands. The light of the long, narrow gas tails was found to be emitted by ions, mainly CO^+, while the diffuse, wider tails consisted of dust particles.

In the 1950's, three major concepts have brought decisive advances to the physics of comets: Oort's model of a distant cometary cloud, Biermann's solar wind theory of tail formation and dynamics, and Whipple's icy conglomerate model of the cometary nucleus. The main ideas of these concepts shall be outlined briefly.

2 Oort's Cloud

The results of statistical investigations of the orbital parameters of nearly parabolic comets led Oort[1] to the interpretation that a cloud of about 10^{11} comets surrounds the planetary system at a distance of some 50 000 a. u.[1], where they are only loosely gravitationally bound to the Sun. Close passages of stars which occur in average intervals of a few million years can perturb the nearly circular orbits of some comets so that they are injected into the inner planetary system. Those which come close to the Sun (perihelia < 4–5 a. u.) become visible as "non-periodic comets". This concept has recently been confirmed and extended by very detailed numerical computations (Marsden et al. [2]). Model calculations have shown that periodic comets with revolution times of several up to 200 years which we observe during multiple perihelion passages (e.g. Comet Encke with a period of 3.3 years and 52 observed apparitions) have probably been captured by the gravitational influence of the massive planets, especially Jupiter, and forced into their present shorter orbital ellipses (Everhart [3]). These comets are frequently called "old" in the sense that they have changed their virgin composition during the repeated exposures to the solar radiation, while the comets coming from Oort's cloud to their first close perihelion passage are "new" in that they have preserved most of their pristine properties.

3 Biermann's Solar Wind Concept

If a comet approaches the Sun, gas evaporating from its surface leads to the formation of the coma. Since the gas drags with it small dust particles which have also been stored in the nucleus, the coma is a mixture of gas molecules and dust particles. Inside about 2 a. u. from the Sun, this material is observed to stream away from the coma in one or two tails (Fig. 1). The separation into two tails is caused by the fact that the gas molecules and the dust particles are influenced by different forces. Relatively early it had been known that the dust tails are shaped by the radiation pressure of the solar light which pushes them away from the Sun with a certain lag angle between the antisolar direction and the orbital motion of the comet. However, the dynamics of the straight, narrow gas tails could not be explained by the light pressure which was too small by one or two orders of magnitude. According to Biermann's theory [4], the solar wind, by interacting with the ionized molecules in the gas tails, is able to accelerate the cometary ions to their observed velocities between 10 and 100 km/s. The small lag angle of only a few degrees between the antisolar direction and the tail axis could be interpreted as an abberrational effect. This angle is given by the ratio of the tangential component of the comet's orbital velocity (typically 30–60 km/s) to the solar wind velocity. Values for the latter up to a few 10^2 km/s deduced from this ratio could later be confirmed by direct measurements from space probes. Furthermore, the solar wind is important for the chemistry of the gas coma, since neutral molecules can be ionized by charge exchange with the solar wind protons. This ionization process must be taken into account especially in the outer parts of the coma, since the inner parts are probably shielded against a penetration of solar wind particles by a contact surface built up at the sunward side of the coma by a pressure balance between the cometary molecules streaming outward and the incoming solar wind flux. Magnetic

---

[1] 1 a. u. (astronomical unit) = $1.496 \cdot 10^8$ km = mean distance Sun — Earth.

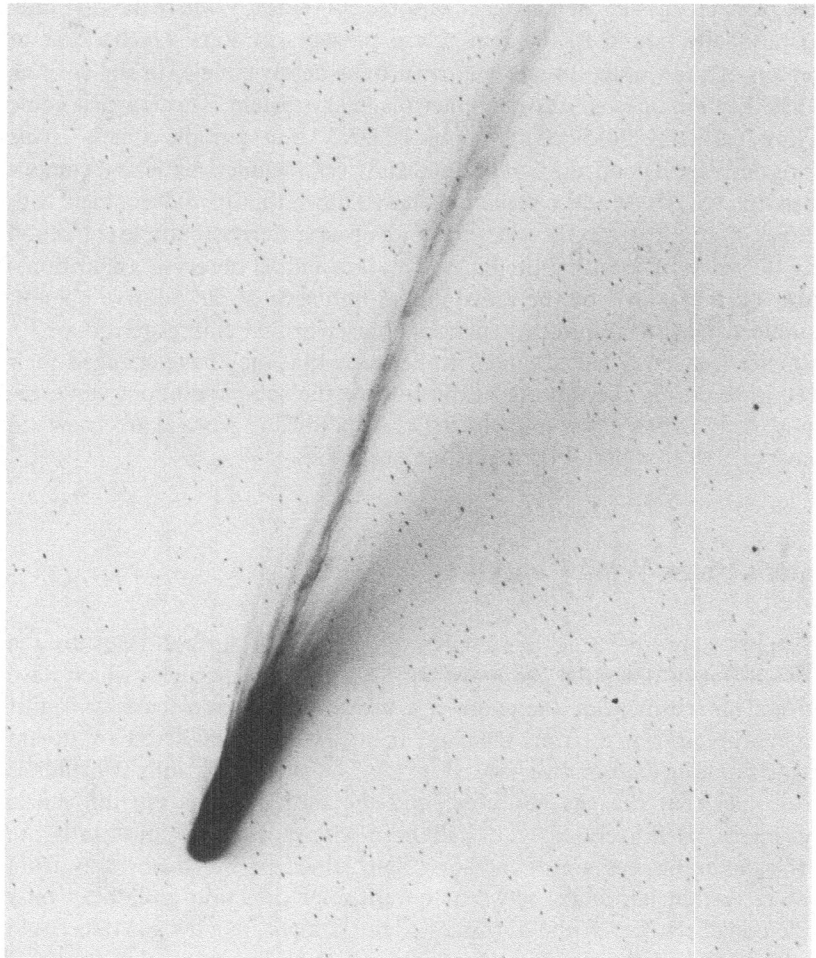

Fig. 1. Comet Mrkos 1957 V with straight gas tail and curved, diffuse dust tail on August 27, 1957. Photogr.: Mt. Wilson and Palomar Observatories

fields embedded in the solar wind play an important role in these complicated interactions which have to be treated with the theory of magneto-hydrodynamics [5].

4 Whipple's Icy Conglomerate Model

After a long debate whether a comet is an extended loose assembly of meteoric bodies kept together by their mutual attraction or a small solid block, Whipple [6] proposed his icy conglomerate model according to which a cometary nucleus is a fluffy, frangile conglomerate of water ice and other frozen gases in which heavier atoms (metals) and dust particles are embedded. Though the "sand bank model" has been strongly defended mainly by Lyttleton, all recent observations and theoretical

Fig. 2. "Sun-grazer" Comet Ikeya Seki 1965 VIII passing through solar corona. Photographed with a coronagrph of Norikura Corona Station in Japan on October 21, 1965

evidence have supported Whipple's "dirty snowball". It is, therefore, generally accepted today. All considerations and estimates for the instrumentation of a space probe to a comet are, for instance, based on this model. However, the final proof is one of the aims of such a mission.

Among the evidence in favor of the "dirty snowball" are observations that many comets — the "Sun-grazers" — have survived perihelion passages very close to the Sun, even through the solar corona, without any noticeable disintegration (Fig. 2). A loose accumulation of numerous small particles would have been completely evaporated. Furthermore, particles of different sizes would have been forced into different orbits by the Poynting-Robertson effect of the solar radiation. Another observation spoke even more strongly against the "sand bank model". Already Encke, at the beginning of last century, has noticed that the comet which had been discovered in 1786 and later was named "Comet Encke" was measurably decreasing its period of 3.3 years. Similar (positive and negative) accelerations were observed subsequently for a number of periodic comets, among them Comet Halley. These findings could be explained with Whipple's model by a repelling rocket effect on the nucleus caused by a mass loss during the active phase when the comet emits gases and dust with velocities of some 100 to 1000 m/s. While a resisting medium, as it had been proposed by the defenders of the "sand bank model", can only decrease the orbital period, the "non-gravitational forces" exerted by the mass ejection can result in an increase as well as in a decrease of the orbital period of a spinning solid nucleus, depending on its sense of rotation with respect to its orbital motion. Rotation periods between some hours and a few days have been determined for some comets

(Whipple [7]). Mass losses of several up to 100 tons per second, depending on the comet's size and distance from the Sun, are necessary to explain the observed changes of the periods, and these numbers agree with the gas and dust production rates derived from observations of the spectra and of the brightness distribution in the comas and tails of some bright comets.

The size of a cometary nucleus cannot be measured directly, since even in the largest telescopes it remains an unresolved point of light. Photometric brightness measurements of comets still far away from the Sun before a radiating halo has formed, together with a phase law and a plausible value for the albedo, yield diameters of the order of 1–20 km (Roemer [8]). Periodic comets are, on the average, smaller than "new" ones, since they lose about 0.1 % of their masses per revolution.

Also the brightness evolution of a comet approaching the Sun is in agreement with Whipple's model. The onset of coma formation at distances as far as 5 a. u. (Jupiter's orbit) points to the existence of frozen gases more volatile than water in the nucleus, such as carbon monoxide, carbon dioxide, methane or ammonia, because the evaporation of water starts only at about 2 a. u. from the Sun. However, the observed dependence of cometary brightness increase with decreasing solar distance — differing considerably from comet to comet — did not fully agree with the model of a simple block of a mixture of water ice and other frozen gases. This increase is much more sudden and the onset of coma formation is later than would be expected if gases of high volatility were the major constituents of the nucleus. The clathrate model proposed by Delsemme and Swings [9] implies that the high volatile gases are embedded in a lattice of water snows, and that these trapped molecules can only escape in larger quantities when the host layer is destroyed by evaporation. If, however, CO_2 and/or CO are present in amounts comparable to that of H_2O as it seems possible from recent calculations of production rates (see Sect. 6), the water ice lattice may not be able to hold these large quantities, and the picture may have to be revised. Experimental studies with ice in a vacuum have simulated this model. The experiments have also repeated the adsorption of gases on a grainy nucleus in a vacuum as it may have happened in an accretion process during the early formation of cometary nuclei (Delsemme and Wenger [10]).

The icy conglomerate model can further explain the different behavior in the brightness evolution of "new" comets and comets having already been exposed to the solar radiation during previous perihelion passages by differences in the outermost layers of the nuclei. "New" comets approaching the Sun generally brighten faster as long as they are still far away from the Sun and much more slowly near perihelion than "older" ones which are not on their first visit to the Sun. An example for this behavior is Comet Kohoutek 1973 XII which remained disappointingly dark after a promising rapid brightness increase after detection. On the other hand, all comets with periods > 25 years show about the same brightness dependence when they recede from the Sun. Whipple proposes as explanation that "new" comets are coated with a layer of very high volatile material which has been gradually formed by an activation by cosmic rays while the comets have been stored for billions of years in Oort's cloud. Once this frosting is evaporated, all comets build up an insulating layer of meteoritic material and — unless their perihelion distances are small — H_2O ice, which is probably very irregular in structure. Returning comets first have to remove this crust before fresh layers of original material can evaporate. Therefore, the evaporation rate increases near perihelion when the crust is blown away by the solar heating (see e.g. Whipple and Huebner [11]).

It must be kept in mind that there are certainly differences in the chemical composition and in the sizes of the comets which add to the picture. For instance, the two comets, Morehouse 1908 III and Humason 1962 VIII, had exceptionally bright CO^+-tails which point to a large content of CO or its parent molecules in the nuclei. Comet Humason was exceptional also because its perihelion distance was as large as 2.1 a. u., a distance beyond which usually no gas tails are visible, and due to its large intrinsic brightness it could be observed for a little over 4 years. Therefore, it may belong to the largest comets observed so far.

5 Molecules of the Nucleus

Until very recently, only secondary products originating from dissociation and ionization of the primary molecules have been identified in cometary spectra. There are different reasons for this. Firstly, most of the primary molecules are rather rapidly destroyed by dissociation and ionization. Only the very stable species can reach larger distances from the nucleus beyond some 10^3 to 10^4 km. Secondly, those molecules which are abundant in interstellar space and could therefore also be constituents of comets (H_2O, CO_2, CO, NH_3, CH_4) do not have strong transitions in the optical spectral region. With the advances of radio astronomy and ultraviolet spectroscopy from rockets and spacecraft, a few probable primary molecules have recently been detected, namely CO and CS in the UV region near 1500 and 2600 Å, and H_2O, HCN, and CH_3CN in the cm and mm wavelength range (for details see Sect. 6). Preceding the detection of neutral water, some previously unidentified red lines could be attributed to the H_2O^+ ion after a comparison with laboratory spectra (Herzberg and Lew [12]). Already earlier, Miller [13] had noticed that the gas tails of some comets which are usually dominated by the blue emission of CO^+ ions also emit a red component which then could not be attributed to any known molecule. After the identification of H_2O^+, these red lines have been found in the spectra of many comets. Furthermore, lines of the CO_2^+ ion were found in the UV and in the optical blue region of three recent comets (see e.g. Rahe [14]). One can therefore assume that CO_2 and/or CO belong to the abundant constituents.

6 The Spectrum of the Coma

6.1 The Visual Spectrum

The wavelength range between roughly 3000 and 8000 Å has been the only source of spectral analysis before the first radio and ultraviolet spectra gave a wealth of new information. The most complete sample of more than 200 spectra from 66 comets can be found in the "Atlas of Representative Cometary Spectra" compiled by Swings and Haser [15]. A detailed interpretation of the optical spectrum was given by Arpigny in 1965 [16]. At that time, the following species had been identified:

Radicals: CN, C_2, C_3, NH, OH, NH_2, CH
Atoms: [OI] (forbidden lines), Na, Fe
Ions: CO^+, N_2^+, CO_2^+ (mainly tail emissions).

The metallic lines, to which in the meantime several other atoms (Ni, Ca, Cr, Mn, Cu, K, Co) have been added, appear only when a comet comes very close to the Sun; the Na emission starts inside about 0.7 a. u., the other lines are weaker and appear later. The lines are generally excited by resonance fluorescence of the incident sunlight, with the exception of the forbidden oxygen lines in the red spectral region. Resonance fluorescence would in this case lead to unreasonably high production rates, due to the small transition probabilities of these lines. Biermann and Trefftz [17] pointed to dissociation processes (photoprocesses or collisional processes) including a dissociative recombination of O_2^+ ions leaving one of the generated oxygen atoms in an excited metastable level of the forbidden transitions. This explanation has an important consequence. Due to the low transition probabilities, every molecule involved in this process can only once contribute to the emission of the forbidden line within approximately 10^5 s while it reaches the borders of the visible coma. Numerical estimates led to production rates for the parent molecules about 2 orders of magnitude higher than those deduced from the lines of the most abundant radicals CN and C_2. This result was the first direct evidence that the radicals which dominate the visual light emission are only minor constituents of the coma. Higher total production rates of the same orders were also postulated from an energy balance between the incident solar radiation on the one hand and the rate of evaporation, reradiation and heating of the nucleus on the other hand (Huebner [18]). The third evidence for higher production rates came from the fluid dynamics model of the cometary dust (Finson and Probstein [19]) yielding dust emission rates of $\sim 10^8$ g/s for Comet Arend-Roland 1957 III and a total gas emission of $\sim 10^{30}$ mol/s, which gives a dust-to-gas mass ratio of the order of 1 (see also Sect. 8).

The assumption that water is a major constituent of the nucleus would explain the high production rate of oxygen deduced from the [OI] lines as a dissociation product of H_2O. In this case, comparable amounts of hydrogen must also be produced, and this consideration led Biermann [20] to postulate the existence of a huge cloud of neutral hydrogen atoms around the coma. Only two years later, this assumption was verified when the two bright comets Tago-Sato-Kosaka 1969 IX and Bennett 1970 II were observed from the two Earth orbiting satellites OAO-2 and OGO-5, which detected a very extended hydrogen atmosphere in the light of the Lyman α resonance line at 1216 Å, (Code et al. [21], Bertaux et al. [22]). In the meantime, also the Balmer line of neutral hydrogen in the red visual spectrum was found.

6.2 The Ultraviolet Spectrum

After the first ultraviolet spectra of two bright comets had been taken, several comets were observed from airplanes, by sounding rockets and by satellites. In all cases, the hydrogen Lyman α emission was the dominating feature. The diameter of the egg-shaped cloud reached more than 10^7 km, which is about 10 times the diameter of the Sun (Fig. 3). The elongated form is caused by an acceleration of the neutral H atoms from the solar radiation pressure in the order of the solar gravitational attraction. The hydrogen cloud was also present around the short periodic comet Encke, though with less intensity and extension, and it seems that it s a general property of all comets. Already in the first low-resolution spectra from OAO-2 and

LYMAN α ISOPHOTES
COMET 1973 XII
JAN. 8.1.1974

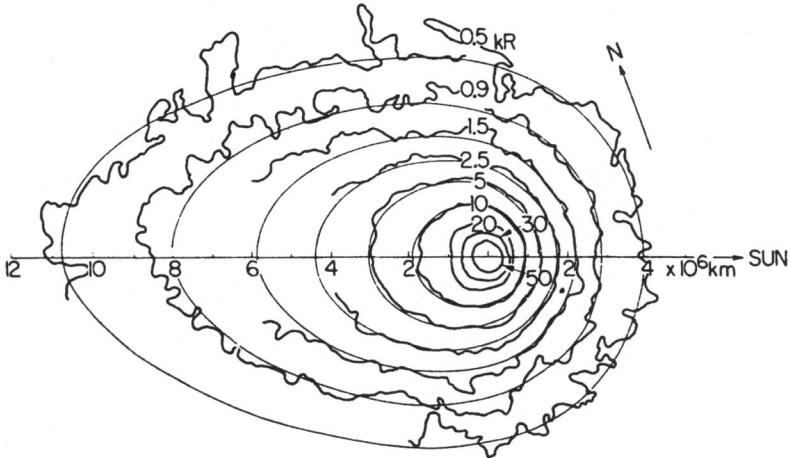

Fig. 3. Lyman α bridhtness isophotes of Comet Kohoutek 1973 XII, intensities in Kilo-Rayleigh, derived from photographs taken with an electronic camera by Carruthers et al. (see Keller [23])

OGO-5, a few other emissions could be identified (Fig. 4). The OH-emissions at 3090 Å which are just barely visible also from the ground turned out to be the strongest lines besides the Lyman α line, and for the first time production rates for this radical could be derived and compared with those for neutral hydrogen. A description of the method by which production rates can be derived from the observed line intensities is e.g. given by Keller [23]. These calculations require the assumption of a certain coma model. The results from several recent comets favor the interpretation that the observed hydrogen and hydroxyl have been formed by dissociation of water. This is roughly demonstrated by the abundances of H and OH given in Table 2. In the case of H_2O dissociation one should expect about twice as many H atoms as OH

Table 1. Production rates for H_2O in 10^{28} molecules/s, reduced to 1 a. u., derived from H and OH productions

	Eccentricity of orbit	Perihel. dist. in a. u.	Prod. rate
Tago-Sato-Kosaka 1969 IX	0.9998	0.473	40
Bennett 1970 II	0.9961	0.538	50
P/Encke	0.847	0.338	0.7
Kohoutek 1973 XII	0.999997	0.142	30
Kobayashi-Berger-Milon 1975 IX	0.9997	0.426	6
West 1976 VI	0.9997	0.197	50
P/Bradfield 1979 X	0.988	0.545	5–10

for authors see Rahe [14]; the eccentricities for the 5 non-periodic comets are corrected for planetary perturbations.

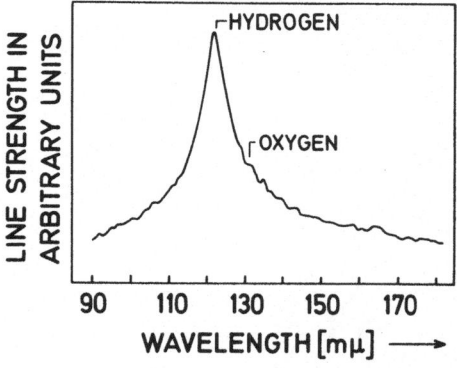

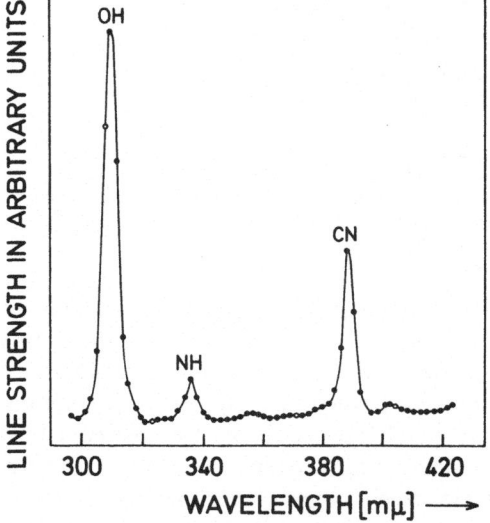

Fig. 4. Ultraviolet spectra of Comet Bennett 1970 II from observations with the OAO-2 satellite (Keller [23])

UV – SPECTRUM OF COMET BENNETT
FROM OAO – 2

radicals, since the H emission comes mainly from distances beyond 10^5 km from the nucleus where OH is already fully dissociated into O and H, while the OH contribution is limited to a region inside 10^5 km. However, in the present state it is not possible to exclude also other parents for the observed H and OH. If, for instance, formic acid (HCOOH) which has been found in the interstellar gas (Table 3) were also present in comets, its dissociation would lead to the formation of H and OH in the same ratio.

Water abundances for 7 comets for which O and OH abundances are now available are given in Table 1. The comets with eccentricities near 1 which have probably been on one of their first close perihelion passages (Kohoutek is the only "new" comet in this list) have much larger H_2O productions than the short periodic comets Encke and Bradfield (periods 3.3 years and ~ 250 years resp.). This result agrees with the icy conglomerate model.

Table 2. Production rates of major coma constituents, in 10^{28} molecules/s reduced to r = 1 au

Species	Bennett 1970 II	Kohoutek 1973 XII	West 1976 VI
H (Lyα)	54	34	46
	65		
	58		
OH (X $^2\Pi_I$)	30	20	20
O (^{3}S) (Resonance line)	6	2.7	23
O (^{1}D) (Forbidden line)	12	1.1	
C (^{3}P)		1.6	6.3
C (^{1}D)			2.7
CO (X $^1\Sigma^+$)			8.5

from Delesemme [25]

A further importance of the ultraviolet spectrum lies in the fact that not only the resonance lines of hydrogen, but also those of the other three basic elements oxygen, carbon and nitrogen fall into the region between 1200 and 1700 Å. While the 1304 Å line of O has already been found in the spectra of Comet Bennett 1970 II (Fig. 4), N has not yet been detected, since it is also blended by the wings of the Ly α line and is probably less abundant than O. The C atom was first identified in Comet Kohoutek 1973 XII. This comet was detected in March 1973 at a solar distance of 5.3 a. u., more than 9 months before its perihelion passage on December 28, 1973 (perihelion distance 0.14 a. u.). Though it did not become as bright as one had anticipated from its early brightness evolution, no other comet has ever been observed in so much detail and by so many astronomers. A whole issue of "Icarus" devoted to this comet appeared at the end of 1974 with the first results, and numerous other publications have followed. Production rates for Kohoutek are compared with the results from two other bright comets, Bennett 1970 II and West 1976 VI in Table 2. Comet West passed through perihelion on February 25, 1976 at a solar distance of 0.2 a. u. and has very successfully been observed by three sounding rocket experiments after perihelion in March 1976. The moderate resolution spectra taken by Feldman and Brune [24] revealed some new constituents, among them atomic sulfur and the molecules CO and CS. Also the C- and O-lines already identified were again present.

The three values in Tab. 2 for the H production in Comet Kohoutek are derived from different observations (for authors see Delsemme [25]). The O-abundances calculated from the forbidden oxygen lines may be regarded as the more reliable ones since the deduction is more straightforward and does not need any model assumption.

While rocket observations can give only snapshots, long-time investigations from satellites have the great advantage of presenting variations of the emissions in space and time. In this respect, the IUE (International Ultraviolet Explorer) satellite which was in orbit when the two bright comets Seargent 1978 XV and Bradfield 1979 X

Table 3. Interstellar molecules (from Winnewisser [28])

2	3	4	5	6	7	8	9	10	11
H_2	H_2O	NH_3							
OH	H_2S								
SO	N_2H^+								
SiO	SO_2								
SiS	HNO								
NO									
NS									
CH^+	HCN	H_2CO	HC_3N	CH_3OH	HC_5N	$HCOOCH_3$	HC_7N	$-$	HC_9N
CH	HNC	$HNCO$	C_4H	CH_3CN	CH_3CCH		$(CH_3)_2O$		
CN	C_2H	H_2CS	H_2CNH	CH_3SH	CH_3NH_2		CH_3CH_2OH		
CO	HCO	$HNCS$	NH_2CN	H_2C_2O	CH_3CHO		CH_3CH_2CN		
CS	HCO^+	C_3N	$HCOOH$	NH_2CHO	H_2CCHCN				
	OCS		CH_4?						

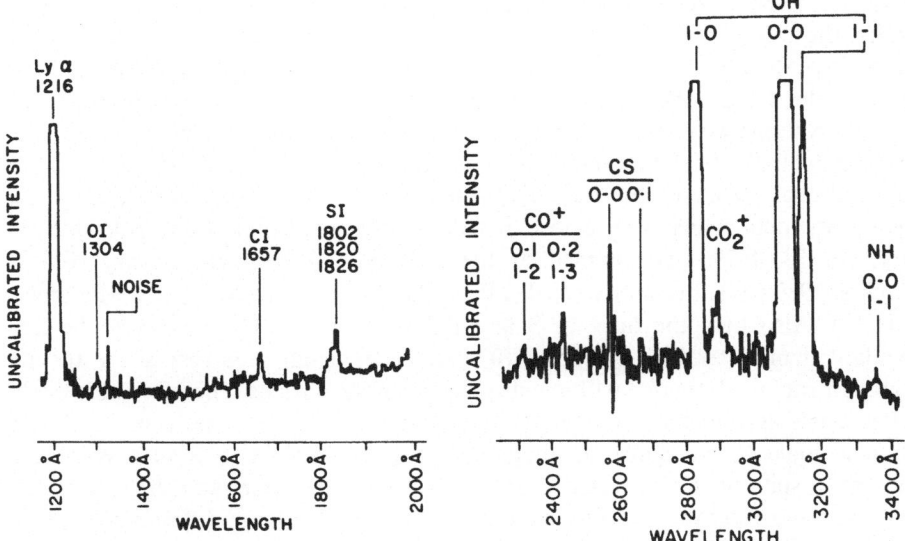

IUE-SPECTRA OF COMET SEARGENT, M=9
LOW DISPERSION, EXPOSURE TIMES 180 AND 165 MIN.

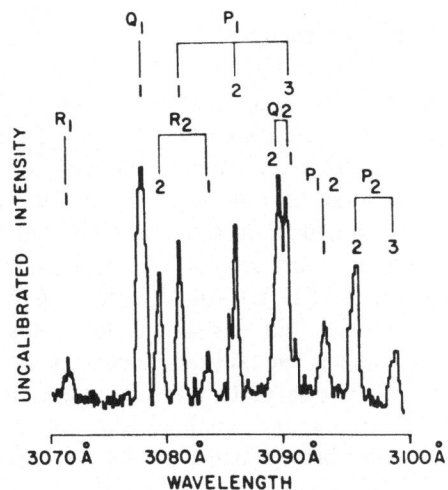

HIGH DISPERSION IUE-SPECTRUM OF RESOLVED OH-BAND

Fig. 5. Ultraviolet spectra of Comet Seargent 1978 XV taken from the IUE satellite. Exposure times for the low-resolution spectra are 180 and 165 min. The high-resolution spectrum shows the rotational structure of the (O,O) OH band (from Jackson, W. M., Icarus 41, 147, 1980)

appeared, deserves special mentioning. The results from the IUE telescope have yielded the most detailed information about comet UV emission obtained so far (Rahe [14]). Already two weeks after Comet Seargent was discovered from New South Wales on October 1, 1978 and four weeks after its perihelion passage, the first observations were made when the comet's distance from the Sun was 0.85 a. u. (Jackson et al. [26]). Both low and high resolution spectra were taken, and in the latter the rotational structure of the OH bands near 3090 Å was resolved (Fig. 5). Comet Bradfield which was detected 3 days after perihelion from Adelaide on December 24, 1979 was observed by the IUE satellite between January 10 and March 3, 1980 while the comet receded from the Sun from 0.7 to 1.5 a. u. (Feldman et al. [27]). Also here the lines of S and CS were visible. CS seemed to be less abundant than water by a factor of 10^{-3}. Its scale length was very short and its brightness rapidly decreased with the distance from the nucleus. Therefore, it is likely that it exists as primary constituent in the nucleus. The UV spectrum of Comet Bradfield looked very similar qualitatively to those of the two comets West and Seargent in spite of the facts that Bradfield is a periodic comet and the two others move on almost parabolic orbits, and that there are also large differences in their perihelion distances as well as in their appearances. This might be taken as an indication that the original composition of all comets is essentially similar and that they have the same history.

6.3 Observations with Radio Telescopes

More than 50 molecules have so far been identified by radio observations of the interstellar matter, most of them containing carbon (Table 3, from Winnewisser [28]). Among them are 7 which have also been found in comets (OH, CH, H_2O, HCN, CH_3CN, CO, CS), the latter two so far only in the ultraviolet spectrum. Since we assume that comets have been formed from interstellar material, radio astronomers have also looked in comets for other species present in interstellar space, for instance ammonia (NH_3), methane (CH_4), formaldehyde (H_2CO) and a few others. Up to now, all results were negative, but this should not be taken as a proof that these molecules are absent in comets. The failure to detect for instance H_2CO might either be due to a large beam dilution or to inadequate excitation (Huebner and Snyder [29]). But certainly ammonia and methane do not belong to the very abundant species. An estimate for the abundance of NH_3 has recently been given by A'Hearn et al. [30]. From a spectrophotometry of Comet West, these authors deduced the NH_2 abundance to be about 3 % of the C_2 abundance. If NH_2 is a dissociation product of NH_3, whose photodissociation rate is about 4 times faster than the rate of ionization, the NH_2 abundance may be a good measure of the production rate of ammonia, which would then be very low.

OH is the only coma constituent which has been found in the ultraviolet, in the optical and in the microwave region. Its two hyperfine transitions at 18 cm (1665 and 1667 MHz) have been identified in 7 recent comets, among them the two periodic comets d'Arrest and Encke. Generally, the line intensities were in agreement with the assumption of ultraviolet pumping by the solar radiation. This model did, however, not fit to the observations of the two periodic comets. Possibly, this is a

Table 4. Spectral identifications in comets

Coma	CN, C_2, C_3, CH, $C^{12}C^{13}$, NH, NH_2	visual
	[OI], OH, Na, Ca, Cr, Mn, Fe, Ni, Cu, K, CO,	
	H, C, O, S, OH, CO, C_2, CS	ultraviolet
	CH_3CN, HCN, H_2O, OH, CH	radio
Gas Tail:	CO^+, CH^+, CO_2^+, N_2^+, OH^+, H_2O^+	visual
	CO^+, CO_2^+, CN^+, C^+	ultraviolet

consequence of smaller amounts of high volatile primordial material in older comets which may cause infrared trapping besides ultraviolet pumping. Production rates from the microwave OH emissions are in good agreement with those derived from the UV spectra. For Comet Meier 1978 XXI, Giguere et al. [31] give a value of 1.3×10^{29} mol/s at 2 a. u. as compared to $2–3 \ 10^{29}$ mol/s at 1 a. u. for other comets (see Table 2). This points to a very productive comet. With an absolute brightness of m = 3.5 this comet whose perihelion distance was 1.1 a. u. belonged indeed to the intrinsically bright ones.

Two stable molecules which can be regarded as primary constituents of the nucleus were identified in the microwave spectrum of Comet Kohoutek, namely HCN at 3.4 mm (Huebner et al. [32]) and CH_3CN at 2.7 mm (Ulich and Conclin [33]). Up to now these identifications could not be repeated in other comets. Production rates are estimated to be some 10^{27} mol/s at 1 a. u., in the range of the visual radicals. The search for these molecules and also for CO which had been detected in the UV spectrum of Comet West was unsuccessful in Comet Bradfield 1978 VII, probably because the production rates of this comet were lower (F. P. Schloerb et al. [34]). Upper limits for the column density of HCN and CH_3CN were less than those derived for Kohoutek, while the upper limit for the CO production was comparable to that inferred from Comet West. Also the very important detection of the 1.35 cm line of H_2O in Comet Bradfield 1974 III by Jackson et al. [35] has not yet been confirmed in other comets.

Though much effort has been made during the last 10 years to detect primary molecules in comets by radio techniques only a few positive results could be achieved (for a review of this very complex subject and for a table of negative results see Snyder [36]). It should be emphasized that these observations are very difficult and that the signals are only marginal in most cases. Furthermore, it is not always easy to coordinate the cooperation with radio astronomers in due time when a bright comet appears. With respect to the importance of these observations, all efforts should however be made to continue the search for molecules in the microwave and radio spectra of future comets.

Table 4 gives a compilation of all species identified so far in comets.

7 Model Calculations of Chemical Abundances

The problem to derive abundance ratios of the primary constituents from the observed secondary products is very complex, since the list of observed coma molecules is still incomplete, and there is a great variety of possible processes in the formation of molecules and ions for which rate constants and cross sections are not always well known. In the inner region of the coma, gas-phase chemistry must be taken into account, since especially ion — molecule reactions are effective and compete with photoprocesses from ionization and dissociation and recombination processes. Further outwards, ionization by charge exchange with the solar wind protons also plays a role. It has not yet been possible to calculate water abundances directly from the weak H_2O microwave emission, and all numbers have hitherto been derived from abundances of H and OH assuming that water is the major component. This has, however, not yet been fully proved. It has already been mentioned that the dissociation of formic acid would also produce H and OH in the same ratio. Oppenheimer [38] mentions the following chemical reactions:

$$O^+ + H_2 \rightarrow OH^+ + H$$

$$OH^+ + H_2 \rightarrow H_2O^+ + H$$

$$H_2O^+ + H_2 \rightarrow H_3O^+ + H$$

$$H_3O^+ + e \rightarrow OH + H_2$$

$$H_3O^+ + e \rightarrow H_2O + H \, .$$

This reaction chain could thus lead to the formation of H_2O^+ ions and also to neutral water molecules and OH radicals from almost any hydrogen-bearing compound in an oxygen-rich nucleus.

The question of the CO_2 and CO abundances has not yet been solved. Delsemme and Combi [37] explained the forbidden red oxygen lines as originating to a major part from dissociation of CO_2 into CO and O. This would yield a CO_2 production of $\sim 10^{29}$ mol/s in Comet Kohoutek. The process would also explain the bulk of the CO observed in Comet West (Table 2). The authors expressed at that time the opinion that the presence of CO_2 in cometary nuclei, though in somewhat lower abundance than water, is more probable than that of CO since the observed amounts of CO could also come from dissociation of CO_2. As model calculations have shown, it seems, however, difficult to understand that the observed abundances of CO^+ ions originate from a pure CO_2 nucleus (see p. 20). Moreover, Feldman and Brune [24]

Table 5. Initial compositions for model calculations of observed coma species

Parent Molecule	Number Densities (10^{13} cm^{-3})					
	1	2	3	4	5	6
H_2O	2.5	2.4	2.2	2.75	2.48	2.90
CO_2	1.5	1.5	1.3	0.0	1.46	0.0
NH_3	0.5	0.5	0.5	0.38	0.56	0.66
CH_4	0.0	0.1	0.5	1.37	0.56	0.66
CO	0.0	0.0	0.0	0.0	0.0	1.72

Compositions 1–4 refer to paper 1 (Giguere and Huebner [39]),
Compositions 5 and 6 refer to paper 2 (Huebner and Giguere [40]).

concluded from the production rates of H_2O and CO deduced for Comet West 1976 VI that these rates are consistent with photodissociation from a source in which CO is 1/3 as abundant as water. The spectroscopic evidence favors CO as the major carbon constituent of that comet.

The most detailed calculations in this field have been carried out by Giguere and Huebner. The method and results are published in two recent papers (Giguere and Huebner [39], Huebner and Giguere [40]). The calculations are based on the Swings and Haser model [41] for the visual emissions and on the model described by Keller [23] for the ultraviolet coma.

The first paper deals with the gas-phase chemistry in one dimension. Starting with four combinations of H_2O, CO_2, NH_3 and CH_4 nuclear abundances (Table 5), rate equations for 441 photoreactions and chemical reactions are solved simultaneously. The second paper starts with a somewhat modified initial composition which varies the abundances of CO_2 and CO. While one combination contains only CO_2 and is similar to one of the first paper, the other assumes a pure CO mixture (Table 5). Abundances for NH_3 and CH_4 are almost the same in both models of paper 2 and about 1/3 of the CO_2 or CO abundance. Furthermore, new rate constants are included, and 25 instead of 3 photodissociative ionization reactions are taken into account. The calculations have shown that these reactions are an important source for the inner coma ions. Two processes which are included shall especially be mentioned in this respect:

$$CO_2 + h \rightarrow CO^+ + O + e. \qquad (1)$$

According to this reaction. CO^+-ions are formed directly from CO_2. On the basis of the processes

$$CO^+ + H_2O \diagup^{H_2O^+ + CO \quad \text{(rate constant } 1.7 \times 10^{-9} \text{ cm}^3 \text{ s}^{-1}}_{\diagdown CHO^+ + OH \quad \text{(rate constant } 8.8 \times 10^{-10} \text{ cm}^3 \text{ s}^{-1}} \qquad (2)$$

a prediction of the CHO^+ abundance is possible from the abundance of the H_2O^+ ions. While the constant expansion velocity used in paper 1 has been replaced by an adiabatic expansion into vacuum at a supersonic velocity of 0.6–0.8 km/s reached asymptotically at a distance of 100 km from the nucleus, velocities from photolysis, chemical reactions, radiation pressure or solar wind interactions are neglected. The results of the second investigation yielded important improvements from the inclusion of the photodissociative ionization reaction (1) in the abundance of CO^+. While it was impossible to account for the observed numbers of CO^+ ions in an average bright comet in any of the four CO_2 models of paper 1, the inclusion of process (1) leads to much higher CO^+ numbers, especially in the inner coma though it is still 6 times slower than the direct photoionization of CO. The CO composition provides, therefore, still much larger CO^+ abundances than the preceding one. In the outer parts beyond 10^4 to 10^5 km, the inclusion of process (1) becomes less effective, and the numbers are more similar. Here, however, the neglection of any sweep-up effects of the ions by solar wind interactions or CO^+ photolysis may make the results somewhat unrealistic (see Table 4 in Huebner and Giguere's paper [40]).

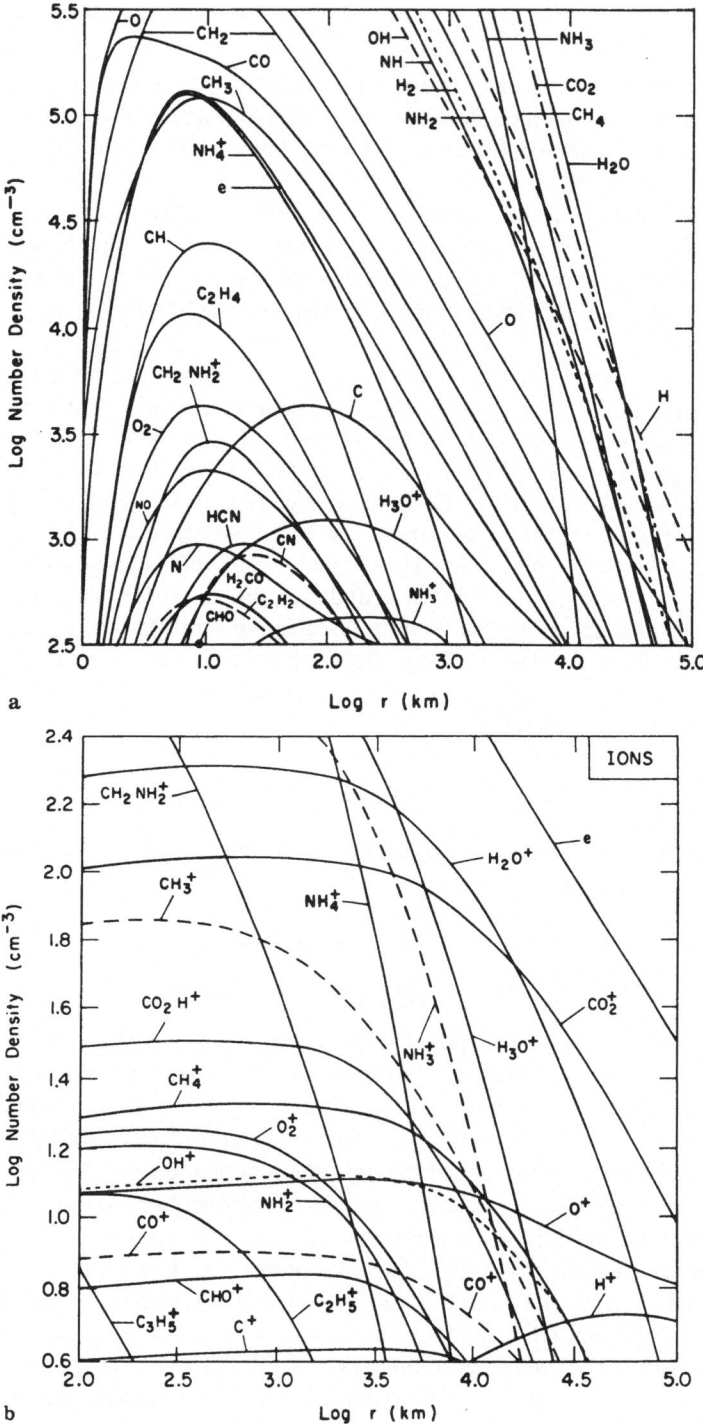

Fig. 6. Results of the model calculations for an initial mixture containing CO_2 and no CO (Compos. 5 of Table 5). Number densities reduced to a solar distance of 1 a. u. are plotted against distances from the nucleus. **a** Inner coma $< 10^5$ km from the nucleus and species with densities > 300 cm^{-3}. **b** Coma between 10^2 and 10^5 km. Ordinates show densities of less abundant constituents (ions) between 4 and 250 cm^{-3}

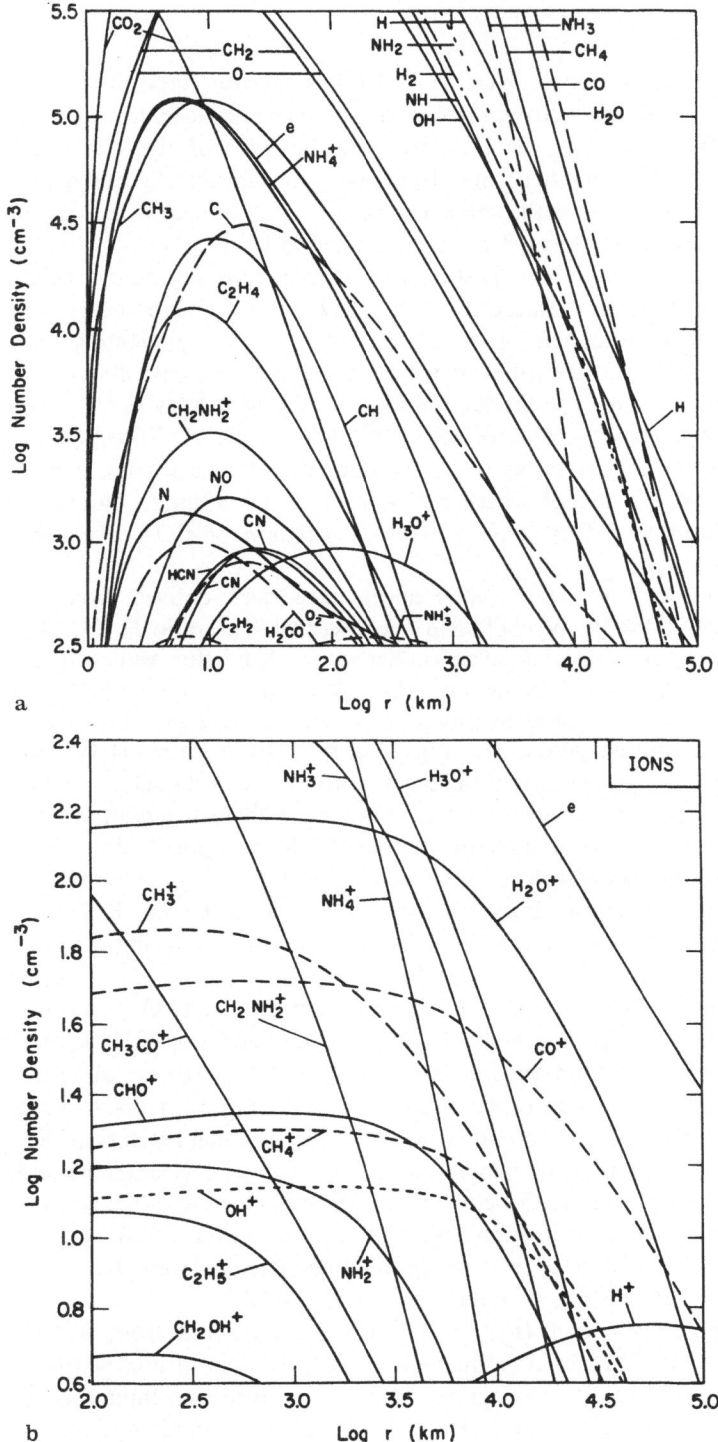

Fig. 7. Same as Fig. 6, for an initial mixture containing CO and no CO_2 (Compos. 6 of Table 5). Note the difference in the CO^+ densities in Figs. 6b and 7b

The CO^+ production in Comet West 1976 VI has recently been studied by Combi and Delsemme [42] from spectrograms covering a range of heliocentric distances between 0.44 and 0.84 a. u. For the first time the influence of the solar wind charged particles has been taken into account by applying the magneto-hydrodynamic theory. Important in this respect is the contact surface in front of the nucleus on the sunward side. Its distance from the nucleus was estimated to $1-3 \times 10^4$ km, depending on the comet's distance from the Sun. This contact surface separates the ions in the inner region, which are hardly influenced by the solar wind, from the outer ones which are accelerated to velocities up to some 100 km/s; thus preventing their detection. Therefore, the major contribution to the observed column densities comes from the region inside 10^4 km. The production rate was found to vary as $r^{-4.6}$ with solar distance and had a mean value of 1.1×10^{27} mol/s at 1 a. u. Direct photo-ionization of the CO molecules seems to be the dominant source for the ions. It cannot be decided by this model whether CO itself is the ultimate parent, or whether it is produced in a first step by photodissociation of CO_2 or another parent.

Next we turn to the H_2O^+ ion. Abundances have been derived from the spectra of Comet Kohoutek by Wyckoff and Wehinger [43]. The ratio H_2O^+/CO^+ was given as 0.03 after a revision of an even smaller value when better values for the lifetimes of the $H_2O^+ A_1$ state became available (Delsemme and Combi [44]). It was not possible to come even close to this ratio by any of the computer runs in Huebner and Giguere's investigations (see Figs. 6 and 7). In all cases, H_2O^+ was overproduced and CO^+ underproduced. The ratios for the CO model are somewhat better due to larger abundances of CO^+, but even there the numbers are always > 1, that means too large by a factor of 50 to 100. Several possibilities which may cause this discrepancy are discussed in the paper. For instance, some of the used rate constants are very uncertain. Furthermore, the nucleus of Comet Kohoutek might have contained much more CO_2 or CO as has been assumed in the initial model compositions.

But besides an underproduction of CO^+ by the calculations, H_2O^+ was also overproduced. This cannot only be caused by a simple overabundance of H_2O, since the model gives even lower production rates for H_2O than those observed. Huebner and Giguere propose as an explanation the presence of an extended source of water vapor caused by icy grains, ejected from the nucleus, which reach considerable distances from the latter in the coma before they evaporate (Delsemme and Miller [45]). This would lower the H_2O^+-content in the inner coma and increase the CO^+ production, since CO^+ is lost by reacting with H_2O (reaction (2)). Moreover, the H_2O abundance has been derived from the production rates of H and OH, and it cannot be excluded that H_2O is not the only parent of H and OH.

Qualitative abundance ratios of H_2O^+/CO^+ have been derived from surface brightnesses of ionized gas tails by a comparison of red sensitive to blue sensitive plates for 13 comets by Miller [46]. The comets could be grouped into three classes, one containing 5 comets with relatively high H_2O^+ brightness, one containing 4 comets with moderate H_2O^+ brightness and a third class with 4 comets which had a very weak red tail. However, there was no noticeable dependence on the comets' orbital parameters or heliocentric distances.

Figs. 6 and 7 show results of the model calculations for the CO_2- and the CO-model.

Figs. b are continuations of Figs. a to lower number densities, starting with a distance of 100 km from the nucleus and containing only ion and electron densities. Note that the abundances of CN and C_2 which give the dominant contribution to the visible coma are very low, the C_2 peak density (near 200 km) is even below the scale of the ordinates. (This peak density was calculated to 0.9 cm^{-3} for the CO_2 model). These numbers illustrate again the extent to which production rates in comets had been underestimated before it became clear that the brightest emissions in the visible coma do not come from the most abundant molecules. Column densities for C_2 and CN derived from the models are, however, somewhat below the observed results. Since the calculated abundances are rather sensitive functions of the heliocentric distance, this discrepancy should not be taken too seriously. C_3 may even be more underproduced by the model, possibly due to uncertain f-values. While the major contributions of C_2 and especially of CN come from the inner coma, the ions are in this model not so much concentrated in a small region. Since solar wind effects which are especially effective outside $\sim 10^4$ km are not considered in this model, ion densities are not very reliable in the outer coma region.

Calculations starting with a more complex initial mixture have recently been carried out by Mitchell et al. [47]. The authors found that the resulting CN- and C_3- abundances agree better with observations in this mixture.

The discussion of the results shows that all such model calculations are extremely dependent on various assumptions (e.g. the ratio of species in the initial composition). Another source for uncertainties lies in the f-values and rate constants which are in many cases not very well known. The calculations give, however, a valuable survey of the processes which may lead to the formation of the observed coma constituents and their relative importances. Also Huebner and Giguere emphasize again the importance to search for other primary molecules, especially to get more reliable estimates whether or in which abundances ammonia and methane are stored in the nucleus.

8 Dust in Comets and Their Relation to Meteorites

So far, we have only discussed the gas production of cometary nuclei. However, about equal amounts (by mass) of small dust particles are emitted from the cometary surface which, after having crossed the coma with velocities of some 100 m/s, are pushed backwards by the radiation pressure of the solar light. With Finson and Probstein's [19] fluid dynamics model, the total production of gas and the dust emission rate can be deduced from the photometry of the coma and the tails. The first results concerning Comet Arend-Roland yielded dust-to-gas ratios varying from 6.2 before perihelion or 0.8 after perihelion. Results on other comets are similar so that generally a ratio of the order of 1 can be assumed. A study of 85 comets ranging from short-periodic to "new" ones did not confirm the former assumption that "new" comets have a larger relative dust production (Donn [48]). The author concluded that the ratio of dust to gas depends more on the grain size of the dust. The data make it less plausible that old, exhausted comets end up as asteroidal bodies.

The sizes of the cometary dust grains vary from less than a micron to probably several centimeters. Infrared observations near 10 μm show the silicate spectral features. In addition, there seems to be a "black" ingredient presumed to be carbon. Due to different accelerations from the solar radiation pressure, the larger particles follow the comet close in its orbit and are more concentrated to the orbital plane. They become sometimes visible in the "anti-tails", narrow spikes which point towards the Sun by an effect of projection when the Earth crosses the comet's orbital plane. Non of the meteorites found so far on Earth seem to be of cometary origin. However, very fluffy micron sized interplanetary dust grains (Brownlee particles) which have been collected by high flying aircraft are possibly cometary debris.

A review of the processes in dust tails has been given by Sekanina [49]. The smaller particles are accelerated to higher velocities, and a considerable percentage can leave the solar system on hyperbolic orbits, especially if the comet is nearly parabolic.

Delsemme [25] has derived the ratio of the oxygen stored in the silicate dust to that in the gases from the dust-to-gas ratios; according to his estimate about 1/3 is contained in the dust. In his model of comet Bennett 1970 II, a combination of this number with the abundances of hydrogen and carbon yields a H/O ratio of 0.7 and a C/O ratio of 0.15 for this comet. His interesting comparison of cometary abundances with abundances in the sun and in carbonaceous chondrites (C I chondrites), which are presumably the most pristine meteorites, shows great similarities in the abundances of these bodies, apart from the fact that hydrogen is depleted by a factor of 5×10^{-4} in comets and $5 \cdot 10^5$ in chondrites with respect to the Sun. During the formation process, the smaller bodies could not keep any free hydrogen. Carbon is depleted by about 1:4 in comets, and the oxygen content in comets and the Sun is nearly the same (Table 6). Both these elements are more depleted in chondrites than in comets. In any other respect, the cometary composition equals that of an interstellar mixture without free hydrogen and helium. Though these results should not be

Table 6. Solar, chondritic and cometary abundances (from Delsemme [25])

	Sun	Comets	C I Chondrites
H	~31 000	15	1.5
C	~12	3	0.7
N	~3	>0.1	0.05
O	~21	21	7.5
Si	1	1	1

overestimated, they are in agreement with a common origin of the Sun, the comets and the chondrites from a presolar mixture 4.6×10^9 years ago and with the assumption that comets belong to the most pristine bodies in the solar system. The origin of the comets is still not known in its details and the subject of various hypotheses, but gradually a picture evolves according to which they may have been formed in the outer cool parts of the presolar nebula or more probably in an adjacent rotating fragment with slightly different initial conditions with respect to its angular momentum and magnetic field (Biermann [50]). In any case, an origin close to the

Sun must be excluded, since all evidence points to extremely low formation temperatures (below ~ 70–100 K).

9 Discussion and Outlook

Photometric observations and the analysis of cometary spectra taken in a wide wavelength range from the ultraviolet to the microwave region have led to a much better qualitative and quantitative picture of cometary chemistry during the last years. The main results are the following:

— The nucleus of a comet is a conglomerate of ices and dust particles of low density (~ 1 g/cm^3). Dimensions of nuclei range roughly from 1 to 20 km. Periodic comets are generally smaller than "new" ones.
— Water ice seems to be the major constituent. Carbon-containing molecules (CO_2, CO) are of comparable, though lower abundances (up to 30%). It is not yet clear whether CO_2 or CO or both are parent molecules of the nucleus. Average production rates for water in bright comets are 10^{29}–10^{30} mol/s in 1 a. u. solar distance.
— Besides H_2O and CO, also HCN, CH_3CN and CS have been identified as probable primary molecules. Production rates for these are estimated to be about 2–3 orders of magnitude lower.
— The search for NH_3, CH_4 and few other molecules in the microwave range was so far negative. This does, however, not exclude their existence. CO_2 has recently been found as ion, but not in its neutral state.
— Atomic hydrogen, carbon and oxygen are visible in the UV spectrum. The hydrogen coma is about 50–100 times larger than the visual coma. Nitrogen has so far not been found.
— The visual neutral coma consists mainly of radicals, CN- and C_2-emissions being the dominant features of the spectrum. Metals (Na, Fe and some others) appear at solar distances below 0.7 a. u.
— Photoprocesses and collisional processes lead to a dissociation and ionization of the parent molecules. Charge exchange with the solar wind is especially effective in the outer parts of the coma while gas-phase chemistry plays a role in the dense central region inside 10^3 to 10^4 km distance from the nucleus.
— The brightness of the gas tail is dominated by the emission of CO^+ ions. Furthermore, H_2O^+ ions and some others are identified in the outer coma and in the tail. The ions are accelerated to some 100 km/s by the solar wind.
— The dust production is of the order of the total gas production (10^7–10^8 g/s at 1 a. u. solar distance). Particle sizes probably range from less than a micron to several centimeters. Infrared observations show the silicate features of the dust. The dynamics of the particles is governed by solar radiation pressure.

"New" comets approaching the Sun for the first time are probably the most pristine objects of the solar system. Their relation to the C I carbonaceous chondrites and the similarity of their compositions suggest that they have come from the same source, the presolar nebula, from which also the Sun and the planets originated. It is, therefore, anticipated that a better quantitative picture of their structure and chemical composition will lead to a better understanding of the formation processes of the solar system. Since many of the open questions may only be solved by "in situ" measurements near a comet, space missions have repeatedly been considered during the last 20 years.

Plans to send a spacecraft to a "new", non-periodic comet are unrealistic so far since the requirements for a high precision pre-calculation of the ephemeris cannot be met during the short time between detection and launch, even if this period would be long enough to allow for the necessary preparations. On the other hand, periodic comets are of lower scientific interest since they have lost their pristine composition during multiple perihelion passages. Comet Halley is the only periodic comet which has still many properties of a non-periodic comet, as e.g. a much larger intrinsic brightness than all other periodic comets and a well-developed gas and dust tail. It has been extensively observed during its last perihelion passage in 1910, which is important for an estimate of the conditions which the spacecraft will meet in the vicinity of the comet. The next apparition of Halley's comet will be in 1986, and so far, missions are planned by Japanese and Russian groups and by the European Space Agency (ESA). Unfortunately, the combined Halley-Tempel 2 — mission planned in the US by the NASA had to be abandoned, and it is not yet clear whether another mission will replace this probe.

The ESA mission which has been named GIOTTO after a painting of the Italian artist showing Comet Halley in the year 1301 is planned as a spin-stabilized spacecraft, to be launched by an ARIANE rocket together with a second spacecraft of the GEOS type. Its 53 kg payload includes an imaging camera, neutral and ion mass spectrometers for the gas and dust, a dust-impact detector, a plasma analyser, a magnetometer and a UV spectrometer. The launch is scheduled for July 10, 1985, and the spacecraft will reach the comet after a flight time of 247 days on March 13, 1986, 5 weeks after perihelion passage when its distance from the Sun will be 0.89 a. u. and that from the Earth 0.98 a. u. (Fig. 8). The relative velocity between comet and probe will be as high as 68 km/s since the comet moves in a retrograde orbit, and at its closest

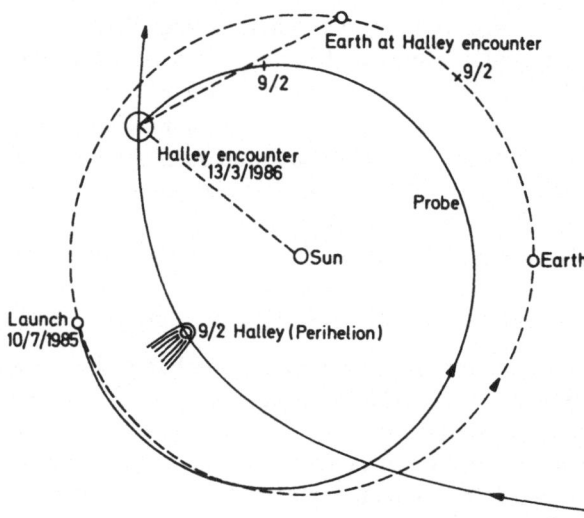

Fig. 8. Trajectory of GIOTTO Halley Space Probe from launch (July 10, 1985) to encounter (March 13, 1986) and orbits of Earth and comet

encounter the probe will be about 1000 km away from the nucleus which will yield a space resolution of ~ 50 m. The probe will pass the comet at its tailward side, and will have crossed the tail in less than half an hour. If this or another of the planned missions will be successful, the measurements will certainly provide us with as many expected and unexpected new results as the previous missions to the Moon and the planets, and they will help us to solve existing problems and raise new ones.

I would like to thank Professor L. Biermann for his interest and for stimulating discussions.

10 References

1. Oort, J. H.: Bull. Astr. Inst. Netherl. *11*, 91 (1950)
2. Marsden, B. G., Sekanina, Z., Everhart, E.: Astron. J. *83*, 64 (1978)
3. Everhart, E.: Astrophys. Lett. *10*, 131 (1972)
4. Biermann, L.: Z. f. Astrophys. *29*, 274 (1951)
5. Biermann, L., Brosowski, B., Schmidt, H. U.: Solar Phys. *1*, 254 (1967)
6. Whipple, F. L.: Astrophys. J. *111*, 375 (1950)
7. Whipple, F. L.: The Moon and the Planets *18*, 343 (1978)
8. Roemer, E.: Astron. J. *73*, 533 (1968)
9. Delsemme, A. H., Swings, P.: Ann. d'Astrophys. *15*, 1 (1952)
10. Delsemme, A. H., Wenger, A.: Pl. Sp. Sc. *18*, 709 (1970)
11. Whipple, F. L., Huebner, W. F.: Ann. Rev. Astron. Astrophys. *14*, 143 (1976)
12. Herzberg, G., Lew, H.: Astron. Astrophys. *31*, 123 (1974)
13. Miller, F. D.: Publ. Astr. Soc. Pacific *67*, 402 (1955) and *70*, 279 (1958)
14. Rahe, J.: Proc. 2nd Europ. IUE Conf. Tübingen, p. XV, 1980
15. Swings, P., Haser, L.: Atlas of Representative Cometary Spectra Louvain, Ceuterick Press 1956
16. Arpigny, C.: Ann. Rev. Astron. Astrophys. *3*, 351 (1965)
17. Biermann, L., Trefftz, E.: Z. f. Astrophys. *59*, 1 (1964)
18. Huebner, W. F.: Z. f. Astrophys. *63*, 22 (1965)
19. Finson, M. L., Probstein, R. F.: Astrophys. J. *154*, 327 and 353 (1968)
20. Biermann, L.: JILA Report Boulder No. *93*, 1968
21. Code, A. D.: paper pres. at the IAU Gen. Ass., 1970
22. Bertaux, J.-L., Blamont, J., Festou, M.: Astron. Astrophys. *25*, 415 (1973)
23. Keller, H. U.: Sp. Sc. Rev. *18*, 641 (1976)
24. Feldman, P. O., Brune, W. H.: Astrophys. J.: Letters *209*, L 45 (1976)
25. Delsemme, A. H.: Proc. IAU Coll. No. 39 Lyon, p. 3 (1977)
26. Jackson, W. M. et al.: Astron. Astrophys. *73*, L 7 (1979)
27. Feldman, P. D. et al.: Nature *286*, 132 (1980)
28. Winnewisser, G.: 2nd DFG Koll. Planetenforschg. p. 184, 1980
29. Huebner, W. F., Snyder, L. E.: Astron. J. *75*, 259 (1970)
30. A'Hearn, M. F., Hanish, R. J., Thurber, C. H.: Astron. J. *85*, 74 (1980)
31. Giguere, P. T., Huebner, W. F., Bania, R. M.: Astron. J. *85*, 1276 (1980)
32. Huebner, W. F., Snyder, L. E., Buhl, D.: Icarus *23*, 580 (1974)
33. Ulich, B. L., Conclin, E. K.: Nature *248*, 121 (1974)
34. Schloerb, F. P., Irvine, W. M., Robinson, S. E.: Icarus *38*, 392 (1979)
35. Jackson, W. M., Clark, T., Donn, B.: Proc. IAU Coll. 25 Greenbelt, p. 272 (1976)
36. Snyder, L. E.: Proc. IAU Coll. 25 Greenbelt, p. 232 (1976)
37. Delsemme, A. H., Combi, M. R.: Astrophys. J. Letters 209, L. 149 (1976)
38. Oppenheimer, M.: Astrophys. J. *196*, 251 (1975)
39. Giguere, P. T., Huebner, W. F.: Astrophys. J. *223*, 638 (1978)
40. Huebner, W. F., Giguere, P. T.: Astrophys. J. *238*, 381 (1980)
41. Haser, L.: Coll. Intern. Univ. Liège *37*, 233 (1966)

42. Combi, M. R., Delsemme, A. H.: Astrophys. J. *238*, 381 (1980)
43. Wyckoff, S., Wehinger, P. A.: Astrophys. J. *204*, 604 (1976)
44. Delsemme, A. H., Combi, M. R.: Astrophys. J. *228*, 330 (1979)
45. Delsemme, A. H., Miller, D. C.: Plan. Space Sc. *18*, 717 (1970)
46. Miller, F. D.: Astron. J. *85*, 468 (1980)
47. Mitchell, G. F., Prasad, S. S., Huntress, W. T.: NASA-JPL prepr. S-22-503 (1981)
48. Donn, B.: IAU Coll. No. 39 Lyon, p. 15, 1977
49. Sekanina, Z.: Proc. IAU Coll. 25 Greenbelt, p. 893, 1976
50. Biermann, L.: Discussion Meeting "Planet. Exploration" of the Royal Soc. London 1980

Marine Manganese Nodules

Dr. Vesna Marchig

Bundesanstalt für Geowissenschaften und Rohstoffe, Hannover, FRG

Table of Contents

1 Introduction

Marine manganese nodules are black to dark brown concretions lying on the sea floor. They are widespread in all the oceans, especially in the areas of low sedimentation rates. Their dimensions vary between less than 1 cm and more than 20 cm (the largest deep-sea nodule that our group had sampled was 24 cm in diameter). The inner structure consists of concentric zones around a nucleus, usually a stone fragment, fossil bone material, or a fragment of an older manganese nodule. The concentric oxide zones have a typical "agate structure" formed by the precipitation of colloids.

The chemical composition of the concentric growth zones around the nucleus shows that they are built of hydrated manganese and iron oxides (predominantly Mn^{4+} and Fe^{3+}) in different ratios. Manganese and iron oxides incorporate some trace metals like Cu, Ni, Co and Zn. The content of Cu + Ni can exceed 2% which makes manganese nodules economically interesting. Smaller manganese nodules usually exhibit a smooth surface; the surface of larger nodules shows a structure, often comparable with cauliflower. The nodules are not very hard (They can easily be scratched with a knife) and readily disintegrate by losing water.

Marine manganese nodules were first sampled during the cruise of HMS Challenger 1872 to 1876. In the following period of time, there was some research about manganese nodules as a geologic curiosity. Within the last twenty years their importance as a future multi-element ore has been realized and this was followed by an intense interdisciplinary research on this field.

2 Chemical Processes Leading to the Accretion of Manganese Nodules

Manganese and iron occupy the 25th and 26th position of the Periodic Table and display a very similar chemical behavior. They can both be counted to the small number of metals which change their oxidation state under the conditions given by the earth. Under reducing conditions, they are soluble as Mn^{2+} and Fe^{2+}, and within an oxidizing environment they are precipitated as highly insoluble hydroxides of Mn^{4+} and Fe^{3+}. This behavior causes the extremely high mobility of manganese and iron within the geochemical cycle.

The other reason for their great mobility is that, as hydroxides, they build very stable colloidal solutions which can be transported over large distances.

Due to the described chemical behavior, manganese and iron hydroxides form widespread precipitates in the form of nodules as well as crusts or disseminated precipitates.

The best known forms of these precipitates are Bog ore, hydrothermal incrustations of manganese and iron, desert harnish, marine manganese nodules and crusts and marine red or brown clay.

The colloidal solutions of manganese and iron hydroxides are, with their very large surfaces, traps for the absorption of different ions from the surroundings during the transport and precipitation. The quantity of absorbed ions depends on the offer of these ions from the surroundings and from the scavenging time. The hydroxide colloids are able to concentrate by means of absorption on their surface ions even from very diluted solutions, if the reaction time is long enough. Because of that, the quickly precipitated hydrothermal manganese crusts generally contain only traces of other metals. Opposite to this, the extremely slowly growing deep-sea manganese nodules can contain enrichments of Cu, Ni, Zn, and Co of a few percent.

In the earth environment, colloidal manganese hydroxide has usually a negative electric charge, scavenging cations from the surroundings. Typical cations which react with the surface of the manganese hydroxide colloid are K^+, Ba^{2+}, Co^{2+}, Ca^{2+}, Li^+, Ni^{2+}, Cu^{2+}, Zn^{2+}, Pb^{2+}, Fe^{3+}.

Colloidal iron hydroxide bears in contrast to that of manganese, usually a positive electric charge on its surface. That is why iron hydroxide scavenges anions from its surroundings as PO_4^{3-}, VO_4^{3-} or AsO_4^{3-}, MoO_4^{2-}.

The attraction between the negatively charged colloid of manganese hydroxide and positively charged colloid of iron hydroxide is also one of the reasons why these hydroxides co-precipitate, building mixed iron-manganese sedimentary ores.

The earth crust contains, on the average, 0.09% manganese and 4.65% iron[30]; hence, the ratio of manganese to iron is about 1:50. By the weathering of rocks manganese and iron are dissolved and transported away in most cases as soluble bicarbonates. Under oxidizing conditions, manganese and iron in weathering solutions are oxidized to Fe^{3+} and Mn^{4+}, respectively, and precipitated as hydroxides. Under reducing conditions, they are redissolved. This process is frequently repeated for several times. In many cases, the precipitates resulting from these processes contain manganese enrichments in comparison with iron; in the extreme case, nearly pure manganese oxide precipitates.

There are several processes causing the enrichment of manganese in comparison with iron:

a) Manganese is reduced by a weaker reducing environment than iron. It dissolves in its reduced form whereas iron is still in its insoluble oxidized form.

b) In the presence of sulfides, iron is bound as a stable sulfide. Manganese sulfide is less stable; actually, it is so instable that it is only found in sedimentary environments formed under strongly reducing conditions.

c) The colloids of manganese hydroxides are more stable than those of iron hydroxide. They can be transported for longer distances, whereas the iron hydroxide already coagulates.

d) Manganese in rock-forming minerals is, on an average, more soluble than iron. This difference in solubility has very little effect on the enrichment of manganese by weathering, but it can become important in the weathering of rocks with hot hydrothermal solutions. The precipitation products from such hydrothermal solutions can sometimes be very pure manganese hydroxides.

Also, the greater stability of iron sulfide could be the reason for the separation of manganese from iron in hydrothermal solutions. These processes have not yet been investigated in more detail.

Thus, another suggestion has been made to explain the separation of manganese and iron in the deep sea[10]. The suggested way of separation can occur only in the radiolarian ooze sedimentary belt in the Central Pacific where radiolarians dissolve in large quantities, and supply the sea-water with Si (the radiolarian tests are opaline SiO_2). From such a sea water containing higher than average quantities of Mn, Fe and Si in solution, iron smectite would precipitate ($Fe_2^{3+}(OH)_2 Si_4O_{10} \times XH_2O$). Iron smectite is a clay mineral which is very resistant and insoluble under deep-sea conditions. As a consequence, manganese would be enriched in the residual solution, from which then manganese nodules would precipitate.

A lot of work will have to be done before we can prove or reject this theory. The first problem is the precipitation of clay minerals directly from the solution. It is nearly impossible to simulate this process in the laboratory, because of the long time needed to initiate the precipitation from relatively diluted solutions. Mostly, clay minerals are formed by recristallization of other silicates which are less stable under weathering conditions. After proving that iron smectite indeed precipitates on the ocean floor, the quantity of this material will have to be determined in order to calculate whether a greater portion of iron can be removed from the sea-water in this way.

The second problem is that the greater portion of iron occurs in the sea-water as colloidal hydroxide. The synthesis of iron smectite should work with iron hydroxide and not only with dissolved iron ions if enough of iron was removed from the sea-water.

Finally, there is the problem why iron smectite is selectively precipitated and not manganese smectite, as manganese has an ionic radius very similar to that of iron, and in most cases both ions can be diadochically exchanged in their compounds ($Fe^{3+} = 0.67$, $Mn^{4+} = 0.52$).

3 Sources of Material for the Growth of Manganese Nodules

There are principally two sources from which deep sea can be supplied with manganese and iron:
3.1 Terrestrial weathering and the following transport by means of rivers to the sea,
3.2 Deep-sea volcanism including the weathering of deep-sea basalt.

Besides these two primary sources, there are two secondary sources which can play an important role in the supply of manganese nodules with metals:
3.3 post-sedimentary remobilization from sediments,
3.4 Dissolution of plankton tests.

3.1 Terrestrial Weathering

The average load of rivers in the world is quite well investigated. We know that rivers transport to the sea 2.2×10^5 t Mn per year in dissolved form and $4 \cdot 10^7$ t Mn per year in suspended form[21, 35]. The deposition of manganese in the *deep* sea is

3.5×10^6 t Mn per year[36]. As we can see from these figures there is much more manganese transported to the sea than sedimented in the deep sea. Even if we assume that 90 % of manganese stays on marginal seas and only 10 % in sediments in the deep sea, we still can deduce all the manganese accumulating in the deep sea only from terrestrial weathering.

All these conclusions have been made on the assumption that the sedimentary load of the rivers is constant through the whole period of accumulation of manganese in the deep sea. Wedepohl[36] believes that the weathering of the earth surface is stronger at present than it has been in the past, due to the pollution and the higher acidity of rain-water caused by pollution.

3.2 Deep-Sea Volcanism

The fact that basaltic volcanism is widespread in the deep sea is established and well proved. This volcanism is present in most areas of the deep sea and especially concentrated on spreading ridges.

Freshly extended basalt heats the sea-water and, as laboratory tests have revealed, the heated saline water can readily leach manganese from the basalt[2, 27].

There is not much known about the quantity of manganese which comes into the deep sea through weathering of deep-sea basalts.

The American submarine "Alvin" was lucky to find on a diving cruise to the Galapogos Rise an active hot spring of metal-bearing water. Galapagos Rise is a place where two basalt plates have recently drifted in opposite directions. In the fracture zone fresh hot lava is coming in direct contact with sea-water. As a result, metal-containing solutions are formed from which the metals precipitate as sulfides by contact with cold sea-water. Such hot metal-bearing springs are highly interesting features, and they will surely help us to understand better the presence of fossil sulfidic ores on the continents. Anyway, they do not seem to influence markedly the chemistry of sea-water or sea sediments. They can always be found in small patches, mostly not larger than a few tens of meters. Therefore, we can conclude that the growth of manganese nodules in the world oceans is independ of such metal-bearing springs.

Part of the volcanism in the deep sea produces volcanic ash — small glassy splinters of basaltic matter, which widely occur in deep-sea sediments. This material (called pyroclastica) weathers relatively quickly because of its instable glassy structure and large surface. How far it contributes to the manganese supply of the deep sea, we cannot say yet, but Beiersdorf[1] considers that it plays a dominant role in the creation of manganese nodule fields.

3.3 Post-Sedimentary Remobilization from Sediments

The manganese accumulating in the sea, no matter from which source it was derived, is either bound to manganese nodules or disseminated in the sediment; the ratio of the manganese in the manganese nodules to the disseminated manganese is 1:6000! The immense quantity of disseminated manganese in the sediment could be the inexhaustible source for the growth of manganese nodules if it could be mobilized.

Near the continental borders in shallow waters there are bright belts of sediments which, because of the high content of organic matter decomposing inside them, are highly reduced (so-called blue muds). From these sediments, most of the manganese is leached and comes into the sea water through diffusion. Iron, although it is also reduced, is bound as sulfide in the sediments, because of the presence of H_2S which is also formed by the decomposition of organic matter.

Our investigations have revealed that in some areas of the deep sea there also occurs remobilization of manganese from the sediment due to the decomposition of organic matter, and that for these areas, the remobilized manganese is the most important factor in the growth of manganese nodules. This process will be explained in detail in the Chapter 9.

3.4 Dissolution of Plankton Tests

Plankton lives in a great quantity in the surface layer of the sea, depending on UV radiation. The predominant part of plankton is composed either of calcium carbonate or opaline SiO_2. After the death, plankton sinks down to the sea bottom. During the sinking different processes are going on. First, the organic matter of the organisms of plankton is oxidized and disappears. During the time of sinking the tests are collecting on their surface different metals from the sea-water. This can occur through adsorption or the metals precipitate as insoluble compounds on the surface.

At about 4000 m water depth, carbonate tests are dissolved, because their solubility depends on pressure (this depth in the ocean is called calcite compensation depth, abbreviated CCD). With the dissolution of tests the metals collected before are again liberated and can contribute to the growth of manganese nodules.

There is not much known yet about the composition of metals collected on the surface of calcite tests. It has only been observed that in the depth where calcite dissolves in great quantities, manganese nodules occur more abundantly and grow especially quickly.

The opaline tests are being dissolved continuously from the sea surface to the bottom, and are further dissolved within the sediment. They seem to be able to supply manganese nodules with Cu and Zn. In the areas opaline tests dissolve in great quantities, the nodules are especially enriched in Cu and Zn[11,23,31].

4 Chemical Composition of Manganese Nodules

Manganese and iron hydroxides are extremely insoluble in sea-water. Maximum concentrations of free ions in sea-water are 0.887×10^{-10} µg Fe/l and 1.09×10^{-6} µg Mn/l, the excess manganese and iron building colloidal hydroxides which sooner or later precipitate.

The ratio of Mn to Fe in manganese nodules therefore depends on the Mn/Fe ratio of the solution from which they precipitate. The ratio of Mn/Fe in the solution is, on its part, governed by the advance of the processes of separation between Fe and Mn as described in Chapter 2.

We know that manganese nodules near the continent have a lower manganese and higher iron content than deep-sea manganese nodules. Iron hydroxide colloids coagulate, as already described, earlier than those of manganese hydroxides. Therefore the first precipitation from the river impact near the continental border should be enriched in iron, and the precipitates more distant from the continental border should have lower iron and higher manganese contents.

We also know that the precipitates from pore water are enriched in manganese, as compared to iron. This is due to the higher mobility of manganese during the process of remobilization from the sediment.

About the manganese-to-iron ratio in the solutions produced by deep-sea weathering of basalt or from deep-sea hydrothermal solutions we do not know very much. The deep-sea precipitates which are evidently hydrothermal can cover the wide scale of compositions from nearly pure iron hydroxide to nearly pure manganese hydroxide.

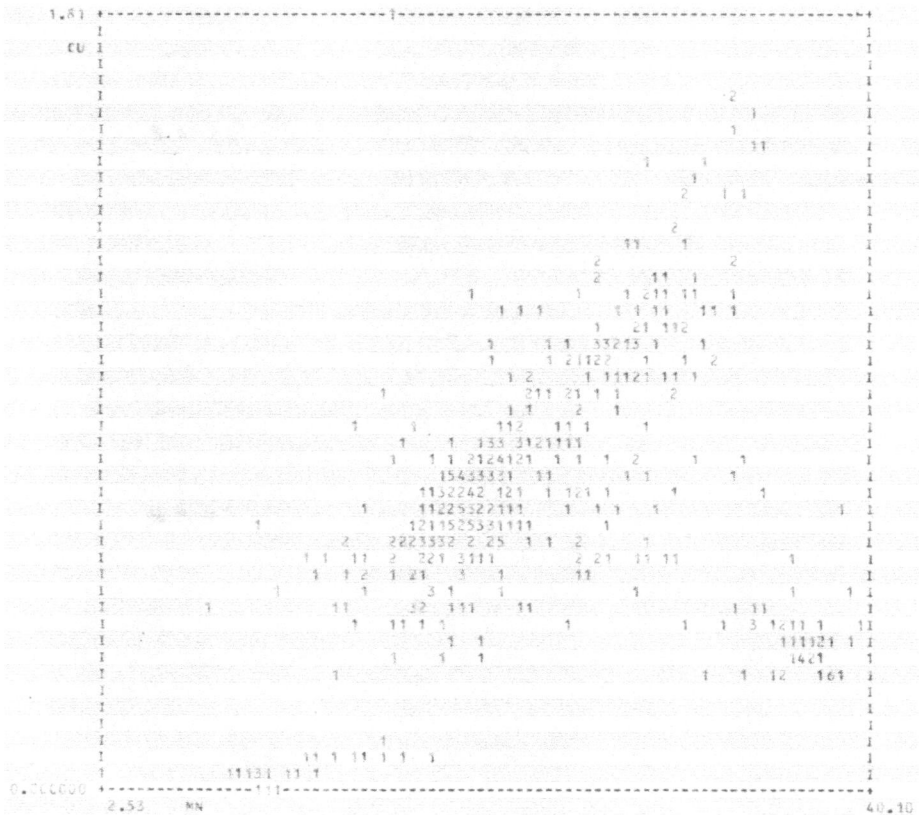

Fig. 1. Cu plotted against Mn for all the manganese nodule analyses performed in our laboratory (mostly manganese nodules from Central Pacific)

During their colloidal phase manganese and iron hydroxides adsorb different ions from sea-water as described in Chapter 2. This process of adsorption can concentrate ions from very diluted solutions. It is dependent on the concentration of ions offered and on time.

Because of the different supply of ions for adsorption in the changing environment, and because of different growth rates of manganese nodules (different time for adsorption) it could be expected that the ratio of adsorbed ions to manganese or to iron hydroxide would be regionally different. There are, anyway, some rules which seem to be valid for the greater part of deep-sea manganese nodules:

Cu, Ni, and Zn are usually bound at constant ratios to manganese hydroxides (Figs. 1–3). The same applies to Mg and Si. Mg and Si are also present in the silicate portion of the nodules. Therefore, correlations of Si and Mg with Mn can be observed only in very "pure" nodules.

Pb and Co are usually bound at constant ratios to iron hydroxide (Figs. 4 and 5).

The third phase of manganese nodules is a silicate phase, containing a nucleus and incorporated sediment. The elements Al, Cr, and Ti occur predominantly in this phase and are thus valuable indicators for the portion of silicates within manganese nodules.

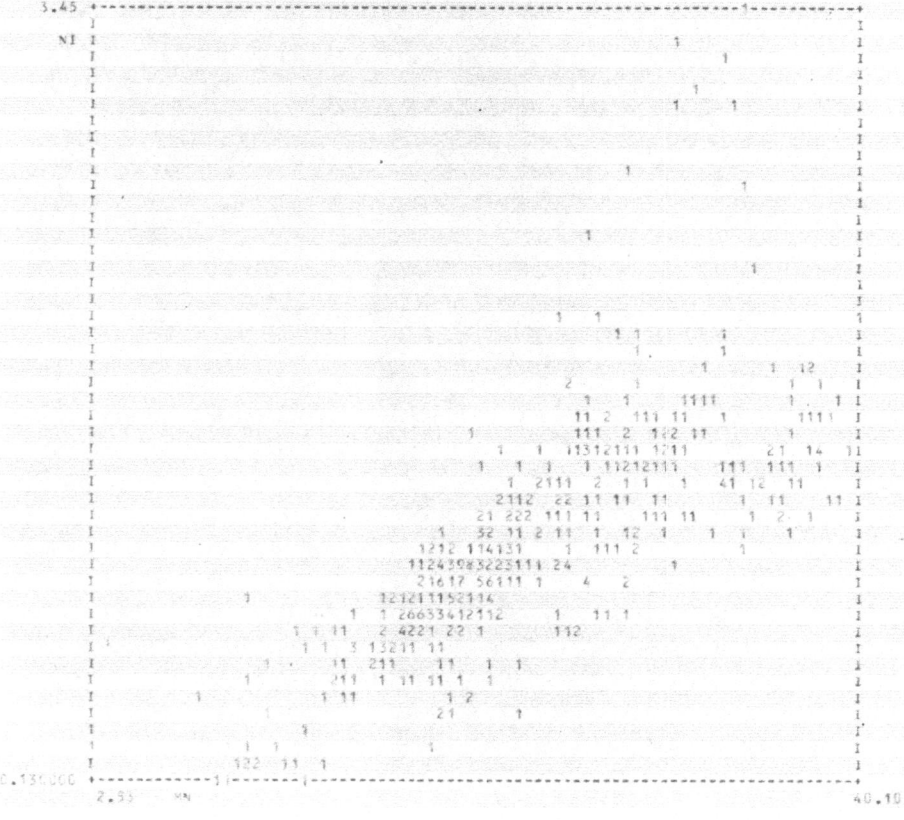

Fig. 2. Ni plotted against Mn for the same samples as in Fig. 1

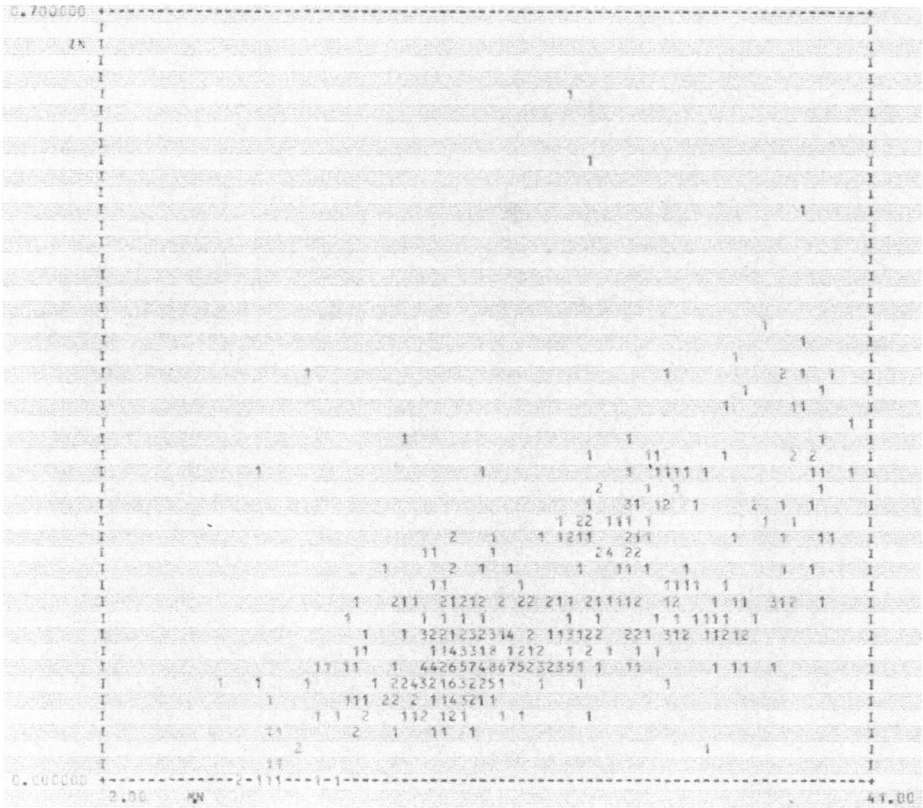

Fig. 3. Zn plotted against Mn for the same samples as in Fig. 1

Sometimes, in the literature the term "hydrogenous part" is used for manganese and iron oxides and the elements incorporated into these oxides within manganese nodules. The silicate portion of manganese nodules is called "terrigenous part".

The major part of elements are enriched in manganese nodules due to adsorption processes on hydroxides. The limited number of analyses does not allow the conclusion in which phase these elements are concentrated. Table 1[5] shows factors of enrichment of a number of elements in nodules, compared with the crustal abundance of these elements.

In a number of cases these data are not based on enough analyses from widespread areas to be statistically proved. It can be expected that, with increasing number of analyses, the average values of some elements in manganese nodules will more or less change.

The precipitated hydroxides age with time — losing water and crystallizing. The adsorbed ions are incorporated into the crystal lattice of newly built minerals due to the possibilities of diadochal incorporation.

There is a lot of confusion in the mineralogical data about manganese nodules. The most frequent manganese mineral in deep-sea manganese nodules is an 10 Å manganate

107

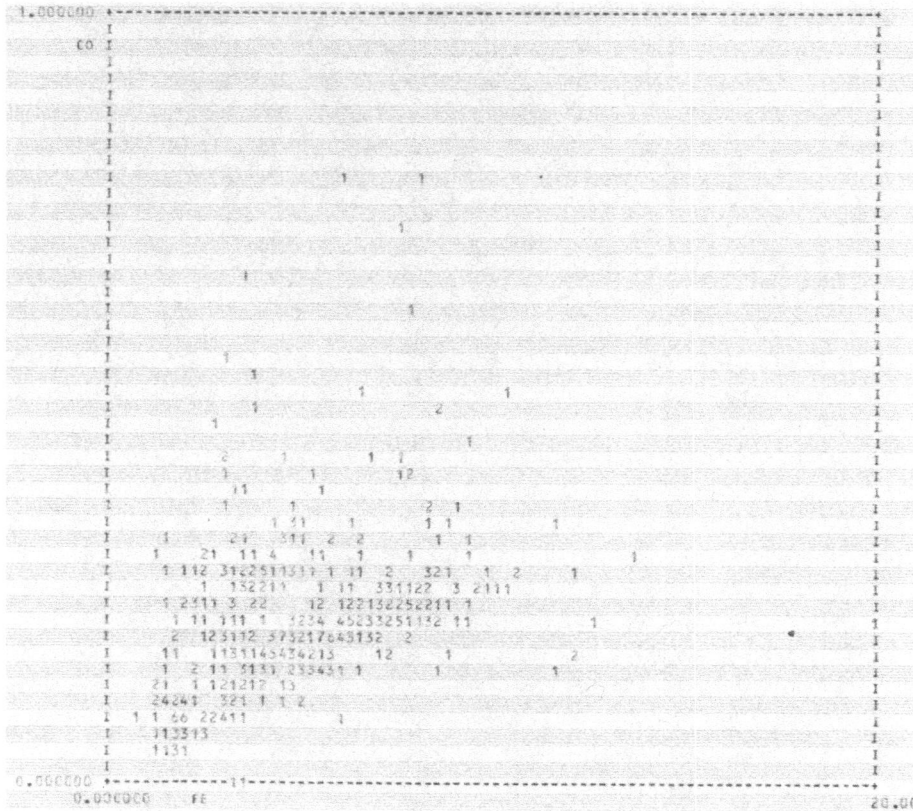

Fig. 4. Co plotted against Fe for the same samples as in Fig. 1

($MnO_2 \times H_2O$ with a crystallographic layer structure, the interval between the layers is 10 Å). This mineral is called "Buserite"[7] or "Todorokite"[4]. It displays strong ion exchange properties especially for transition metals. The ion exchange decreases in following order: Cu, Co, Zn, Ni, Na, Ca, Mg[7].

The dehydrated variety of 10 Å manganates is the 7 Å manganate Birnessite. It can occur as primary mineral in manganese nodules but is also frequently built secondarily after drying and preparation of manganese nodules. This 7 Å manganate has significantly weaker ion exchange properties than the 10 Å manganate.

Sometimes, Birnessite is found with a disordered cristallization structure and does not reveal a cristalline structure by X-ray analysis. To make the confusion even larger, different names have been given to this dissordered Birnessite. Thus, it is called "δ MnO_2$" or "Vernadite".

The iron hydroxide in manganese nodules cristallizes possibly as the mineral "Goethite". The crystallization of Goethite is presumed to occur secondarily by drying and in the preparation of the manganese nodules. The primary iron hydroxide is amorphous.

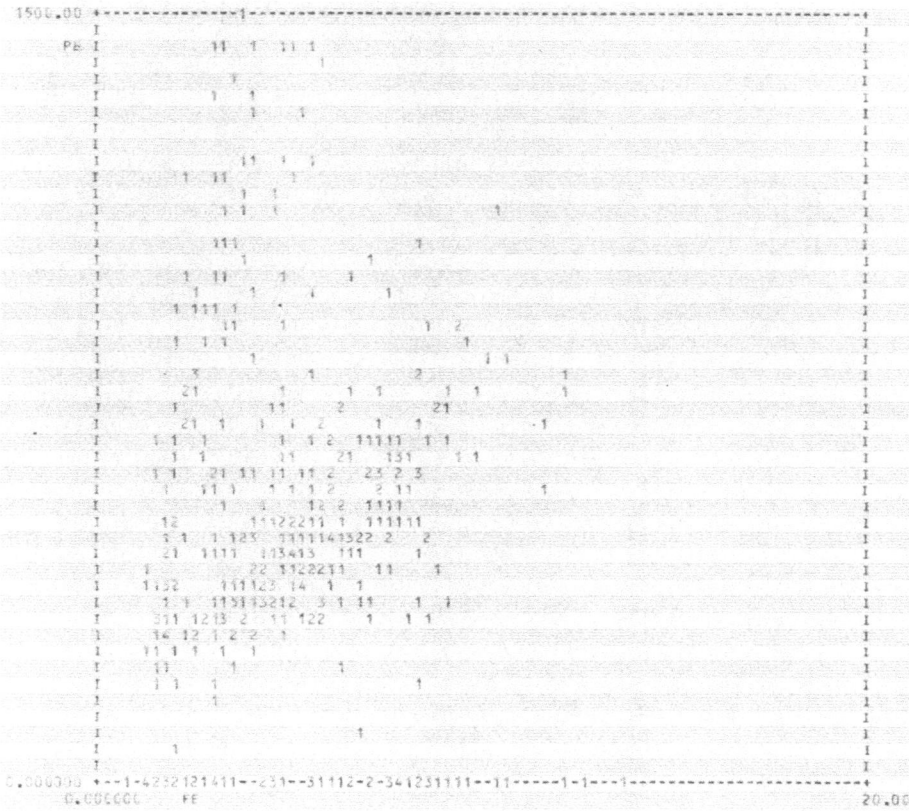

Fig. 5. Pb plotted against Fe for the same samples as Fig. 1

Cobalt is mostly incorporated into the iron phase of manganese nodules. However, theoretically, it would suit much better in the manganese phase which has experimentally been confirmed by Giovanoli and Brütsch[7]. There is a theory, not yet proved by experiments, stating that Co^{2+} is oxidized to Co^{3+} in sea-water. This process should be catalyzed by $Fe(OH)_3$. The final product of oxidation, $Co(OH)_3$, forms a solid solution with $Fe(OH)_3$[3]. The theory does not include a discussion about the possibilities of diadochic incorporation of cobalt after cristallization of iron hydroxide.

Lead, which is also bound to the iron phase has an ionic radius much too large for the diadochic change with Fe^{3+}.

According to recent knowledge about manganese nodules it seems that all the processes occurring with manganese nodules after the precipitation of hydroxides do not change the average composition of manganese nodules in spite of the ion exchange qualities of some new minerals. The only change in chemical composition from older to younger materials which can be observed involves a loss of a few percent of water.

Table 1. World oceanic average of elements in manganese nodules and enrichment factor for each element in nodules compared to crustal aboundance (from Cronan[5])

	element	world ocean average	crustal abundance	enrichment factor
enriched	Tl	0.0129	0.000045	286.66
in nodules relative	Mo	0.0412	0.00015	274.66
to their crustal	Mn	16.174	0.095	170.25
abundances	Co	0.2987	0.0025	119.48
	Ag	0.0006	0.000007	85.71
	Ir	0.935^{-6}	0.132^{-7}	70.83
	Pb	0.0867	0.00125	69.36
	Ni	0.4888	0.0075	65.17
	Bi	0.0008	0.000017	47.05
	Cu	0.2561	0.0055	46.56
	W	0.006	0.00015	40.00
	Cd	0.00079	0.00002	39.50
	B	0.0277	0.0010	27.7
	Sn	0.00027	0.00002	13.50
	Xb	0.0031	0.0003	10.33
	Zn	0.0710	0.007	10.14
	Y	0.031	0.0033	9.39
	Hg	0.50^{-4}	0.80^{-5}	6.25
	La	0.016	0.0030	5.33
	Ba	0.2012	0.0425	4.73
	V	0.0558	0.0135	4.13
	Zr	0.0648	0.0165	3.92
	Fe	15.608	5.63	2.77
	Sr	0.0825	0.0375	2.20
	P	0.2244	0.105	2.13
?	Ti	0.6424	0.570	1.13
	Pu	0.553^{-6}	0.665^{-6}	0.832
	Na	1.9409	2.36	0.822
	Mg	1.8234	2.33	0.782
	Ga	0.001	0.0015	0.666
	Au	0.248^{-6}	0.400^{-6}	0.62
	Ca	2.5348	4.15	0.610
depleted	Sc	0.00097	0.0022	0.441
in nodules relative	Al	3.0981	8.23	0.376
to their crustal	K	0.6427	2.09	0.307
abundances	Si	8.624	28.15	0.306
	Cr	0.0014	0.01	0.14

5 Manganese Nodule Nuclei

Manganese nodules result from the precipitation of manganese and iron hydroxides from sea-water or pore-water around an offered nucleus. Nuclei available in the deep sea are pieces of volcanic material (balsalt, pumice) (Fig. 6) or apatitic fossils (shark teeth (Fig. 7), auditory canals of whales).

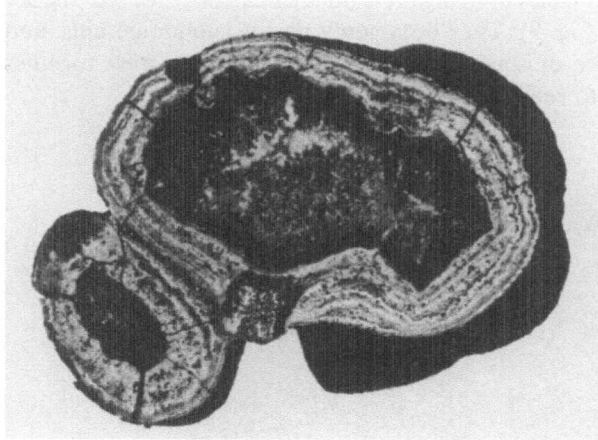

Fig. 6. Cross section of a poly-nucleated manganese nodule. Both parts of the nodule have a basaltic nucleus. In the larger nucleus, dendritic impregnations of manganese in basalt are visible.
Photo: H. Silber BGR. Size of this nodule: $2.7 \times 2.1 \times 1.7$ cm

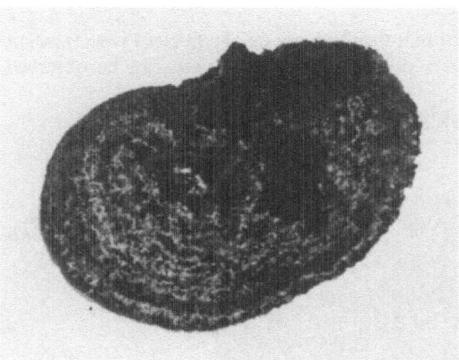

Fig. 7. Cross section of a small manganese nodule with a shark tooth as one of the nuclei (V shape).
Photo: H. Silber, BGR. Size of this nodule: $1.3 \times 1.2 \times 0.9$ cm

Precipitated hydroxides age with time, losing a fraction of incorporated water. Their volume decreases because of dehydratation; as a consequence, they scrimp and crac until total disintegration occurs. The disintegrated parts of older nodules are the best nuclei for the next generation of nodules, because already precipitated manganese and iron hydroxides catalyze further precipitation (Fig. 8).

We suppose that this process of disintegration of older nodules is important for the quantity of nodules growing on the manganese nodule field. The fragments of disintegrated nodules offer for further precipitation a surface which is orders of magnitude larger than that of non-disintegrated nodules. The larger portion of hydroxides precipitating on a manganese nodule field settles in the disseminated form. Only that portion of hydroxides which finds a suitable surface contributes to the formation of manganese nodules. The large surface of disintegrated nodules shifts the ratio of disseminated hydroxides to those in nodules in favor of nodules. In this way, rich manganese nodule fields are formed in positions where otherwise only a small number or large nodules would grow.

In some nodule areas, special facies of manganese nodules can be observed (the so-called polynucleated nodules, Fig. 9). They occur only on the submarine hills, and seem to be built of a number of small maganese nodules which grew together because they were lying close to each other.

Fig. 8. Cross section of a large manganese nodule from a deep-sea plain. The nucleus is a fragment of an older nodule. In the surrounding oxide layers different types of growth can be observed (dendritic and laminated).

Photo: H. Silber, BGR. Size of this nodule: 7 × 5.5 × 3 cm

Fig. 9. Cross section of a typical polynucleated manganese nodule. Basalt fragments are coated and cemented together with oxyde layers.

Photo: H. Silber, BGR. Size of this nodule: 3.8 × 2.2 × 1.9 cm

Submarine currents are stronger on the hill than in the plain because of the reduced current profile. The softer part of the sediment can be partially drifted away with stronger currents. The larger particles as volcanic fragments cannot be drifted away and concentrate on the submarine hills.

The polynucleation of maganese nodules from submarine hills is probably caused by the high number of offered nuclei (volcanic fragments) lying close to each other.

Sometimes, the precipitates do not form maganese nodules but manganese crusts. Usually, manganese crusts are growing on the hard substrate, i.e. on the fresh basalt not yet covered with sediment, or at positions where hard rock is in direct contact with sea-water because strong currents make sedimentary deposition impossible. In such cases crusts are simply "overdimensioned" nodules.

In rare cases, one can find crusts on the sediment at positions where usually nodules would grow. We do not know why in such position sometimes crusts and sometimes nodules grow. One possible explanantion could be as follows: crust is a primary form built from undisturbed precipitation, because of disturbances of precipitation nodules instead of crusts are growing. The disturbances have their origin in the bioturbation of sediments. According to this theory, nodules would grow on the areas where bioturbation is strong (and this is nearly everywhere in the deep sea), and crusts would grow on the rarely occuring sediments not disturbed by bioturbation.

Parallel observations have been made in the case of manganese bog ore. This bog ore precipitates as crust in the soil. The parts of this bog ore crust are disturbed by tree roots: on these parts the precipitate has the form of nodules (Ortlam personal communication).

6 Distribution of Nodules on the Deep-Sea Floor

Deep-sea manganese nodules can widely be observed in all oceans. The majority of them is located on the sediment surface so that the upper side is in contact with the sea-water and the lower side in the nearly liquid mud of the sediment/water interface. Only a small part of nodules gets buried during their growth.

The nodules can be found preferentially in areas with very low sedimentation rates.

The three oceans of the world have different average sedimentation rates. The Atlantic Ocean, not very large, with a great number of rivers transporting their sediments to it, has the highest sedimentation rates. The Pacific, the largest of the oceans, obtaining sediment from a small· number of rivers only, and with deep trenches on its borders acting as sediment traps, has in its deep sea areas the lowest sedimentation rates.

The Indian Ocean has sedimentation rates ranging between those of the Pacific and the Atlantic Ocean.

So it is not surprising that the largest quantities of manganese nodules can be found in the Pacific Ocean.

Figs. 10 and 11 show the distributions of manganese nodules and of the sedimentation rates in the world oceans. From the similarity of these two distributions it can be concluded that the main factor regulating the distribution of manganese nodules is the sedimentation rate.

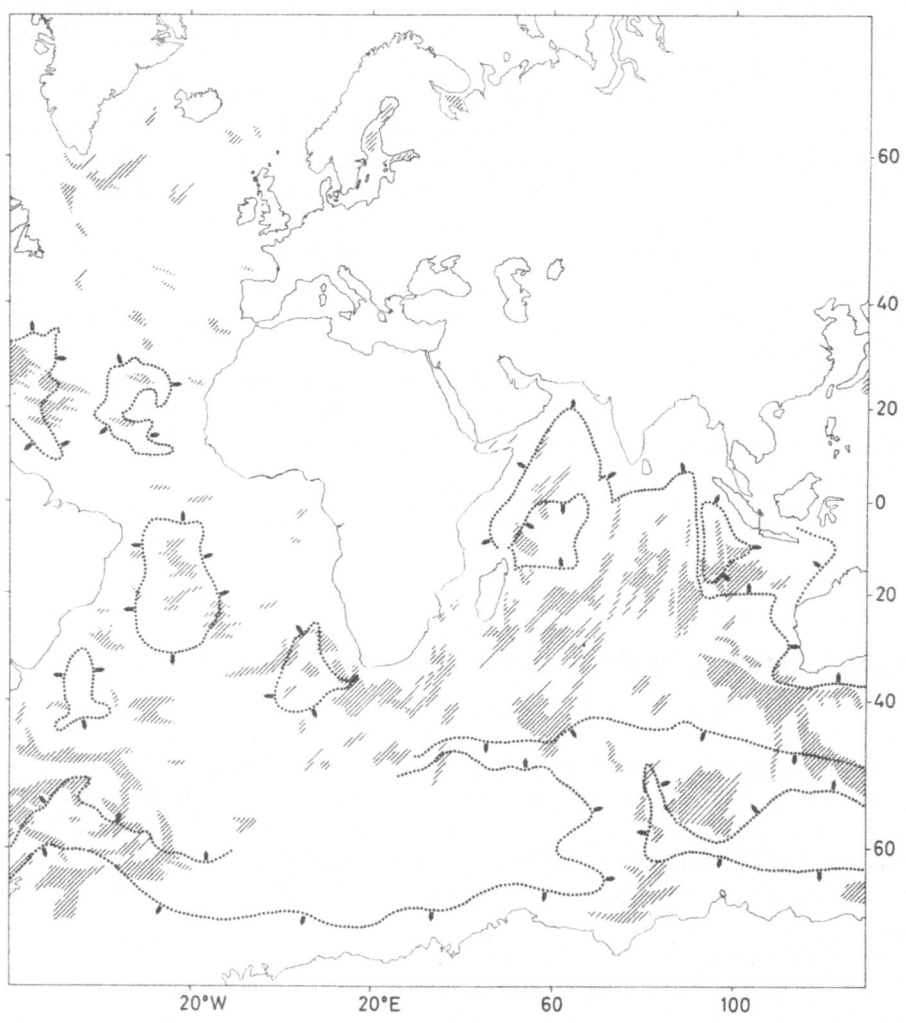

Fig. 10. Map of the Atlantic and Indian Ocean with manganese nodule fields (hatched) and limits of sedimentation rate 30 mm/1000 years (pointed). The arrows indicate the direction of higher sedimentation rates. The data are taken from Lisitzin[20] and Dreyfus et al.[6]. They are inadequate because measurements on the ocean floor are scarce. The map only illustrates that oceans with lower sedimentation rates contain larger fields of manganese nodules and vice versa

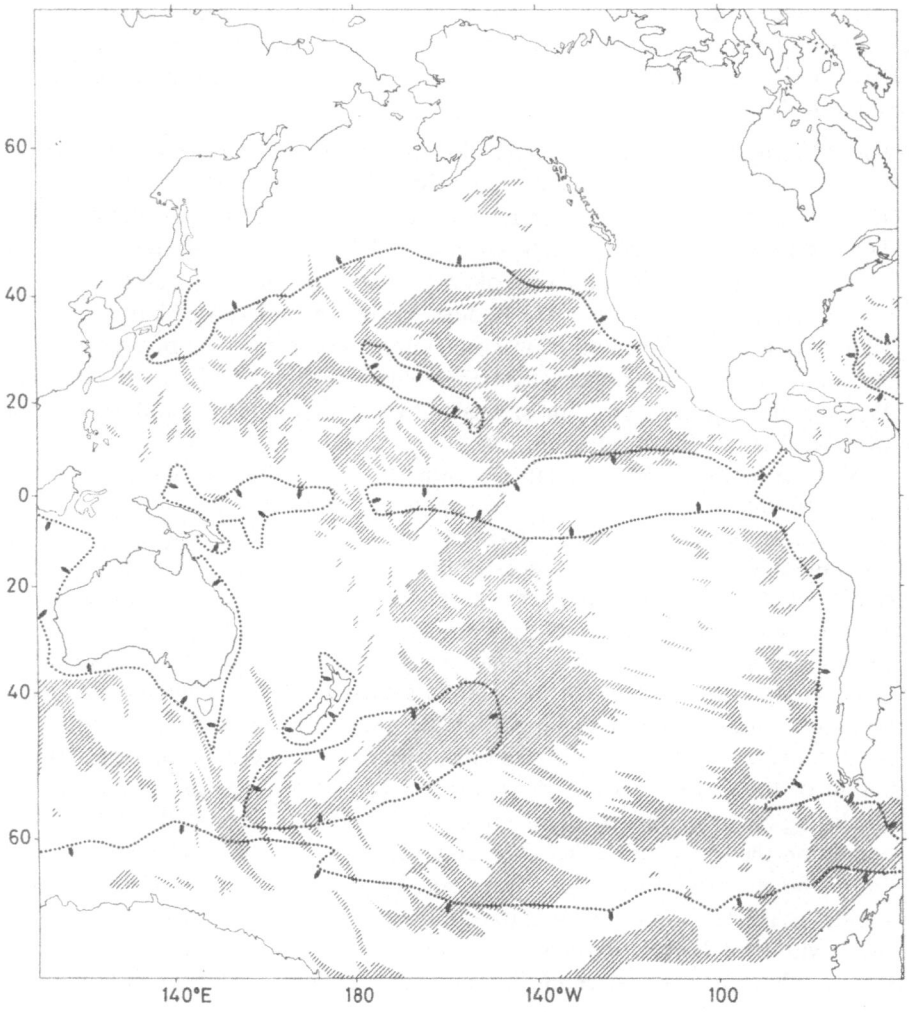

Fig. 11. Map of the Pacific Ocean with manganese nodule fields and limits of sedimentation rates (30 mm/1000 J). For further explanations see Fig. 10

7 Discrepancy between Growth Rate of Manganese Nodules and Sedimentation Rate

The growth rates of deep-sea manganese nodules lie between 4–9 mm in one million years[13]. These growth rates are much lower than the sedimentation rate of the sediment on which the manganese nodules are located. Thus, they should actually get buried within a relatively short period of time.

There must be an explanation why the nodules, in spite of this fact, are located on the surface of the sediment during all the time of their growth. ·

This problem has not yet been solved; there are three proposals which try to explain this dicrepancy:

a) The sediment is not only deposited but also eroded. Through erosion of the sediment the already burried nodules again reach the sediment surface[33].

There cannot be any doubt, that a great deal of erosions occur in the deep sea. The sediment is extremely finely grained and the bottom currents are strong enough to transport it from one place to another. There are even some sea bottom photographs showing the nodules in different phases of burying. Other sediments where sedimentation was not interrupted by phases of erosion were also found, and these sediments also contained manganese nodules on their tops. Thus, erosion could account for one part of the nodules located on the top of the sediment but not for all of them.

b) From time to time. sediment-eating worms lift the nodules or even turn them over, and thus free them from the sediments on their tops.

A strong bioturbation is found regularly in the deep-sea sediments containing manganese nodules[34]. Worm traces of up to 5 cm diameter can be observed in nearly all the sediment samples from the maganese nodule areas. Because the deep-sea sediments contain very little organic matter which serves as food for the worms (0.1–0.3% C_{org}), the worms have to transport large quantities of sediment through their digestive system in order to survive. It can therefore be expected that from time to time these worms lift the nodules lying on their way through the sediment. Radiographies of sediment samples make visible the dense distribution of the worm traces in the sediments (Fig. 12). Glasby[8] calculated that one lifting of the nodule by the worm every 30,000 years is enough to retain this nodule on the sediment surface. We observed on board ship that most nodules lie on the sediment surface in a very soft unconsolidated mud and very little strength is needed to move them. Some of the nodules, usually the very large ones, reach with their bottom parts a harder more consolidated sediment, and can be moved only with greater strength. It is not imaginable that sediment-eating worms can move or even turn these "anchored" nodules.

c) French scientists[17–19] have stated that measurements of growth rates of manganese nodules (performed through measurement of uranium-radioactive decay products) are principally wrong, because radioactive isotopes are not enclosed in the manganese nodule during its growth but become incorporated with adsorption processes on the already accreted material. The manganese nodules grow actually very quickly within few thousands of years, and thus the problem of being buried with slow sedimentation does not exist any more. They have proved their hypothesis by experiments on a few nodules containing young nucleus material.

Heye[15] objects to this theory that measurements of the growth rates of radioactive decay products have been carried out with two different pairs of elements, and that it is improbable that both pairs of metals would have the same adsorption behavior to pretend the identical growth rates.

Very quickly growing nodules (growth by means of supply with concentrated solutions generated through hot weathering of basalt) have been found in the deep sea.

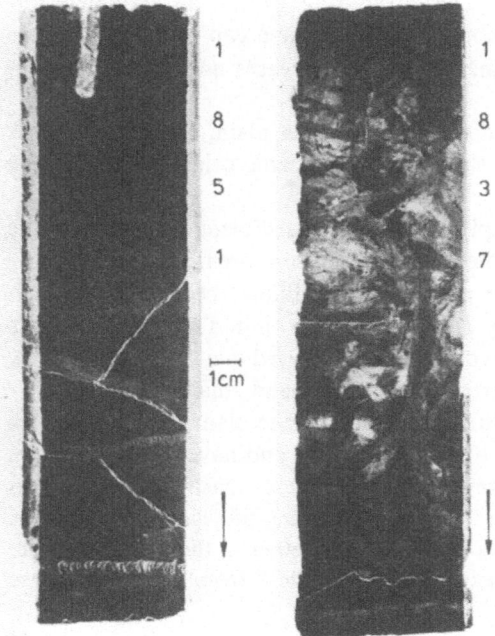

Fig. 12. Radiography of two cores with worm traces (without magnification). Photo: H. Karmann BGR

They are however in minority and cannot explain the large quantity of nodules on the ocean floor.

8 Manganese Nodule Fields on the Radiolarian Ooze Belt in the Central Pacific

In this chapter the manganese nodule fields on the radiolarian ooze in the Central Pacific will be described in detail. These manganese nodules fields are the richest yet known in the world so that they have been the subject of a great deal of research. Probably, the first raising of manganese nodules as ore will take place in this area.

The radiolarian ooze belt is situated in the North Central Pacific between the Clarion and Clipperton fracture zones, about 10°–20 °N and 120°–160 °W. The reason why the radiolarians are sedimented there is a corresponding oceanic current in the surface water, in which radiolarians have optimal conditions for their growth.

The radiolarians settle down on the sea floor together with terrestrial clay particles creating the "radiolarian ooze". The radiolarian tests have a very fragile neat structure (Fig. 13) with hollows inside. These tests thus exhibit an extremely low volumetric weight. Already a small amount of radiolarians can markedly change the behavior of the sediment; its volumetric weight decreases and its porosity increases with the radiolarian content.

The area where the sediments settle down is a deep sea plain, somewhat deeper than 5000 m, interrupted by numerous steep hills of volcanic origin, usually a few hundred of meters high.

The sediment mostly settles in the plain; a thinner sediment cover is on the slopes of the hills and the steep parts of the hills and their tops are free of sediment.

The top of the sediment cover in the plains is nearly liquid because of its high water content. On the slopes of the hills this nearly liquid sediment top is missing; it cannot be formed because it would flow downward due to its consistence.

The whole sediment is strongly bioturbated by worms and smaller fauna.

The large quantity of manganese nodules is located on the plain sediment (on an average 13 kg/m^2). They measure up to 10 cm in diameter and have a rough surface similar to cauliflower. These nodules are rich in manganese. On the deep sea hills there are less manganese nodules (on an average 0.4 kg/m^2), they are smaller, often formed from several even smaller nodules which grew together in the course of time. They are called "polynodules" in contrast to "mononodules" from the plain. These nodules are rich in iron.

The sea water in this area is enriched in some metals compared with the world average sea water. Especially, iron and manganese are present in higher quantities, the iron quantity is several times larger than the quantity of manganese. The patchy distribution of this enrichment seems to be caused by the colloidal stage of iron and manganese hydroxides[9, 37].

This iron and manganese hydroxides from sea-water are precipitated at the sea bottom in very different forms. If there is any solid nucleus available for preci-

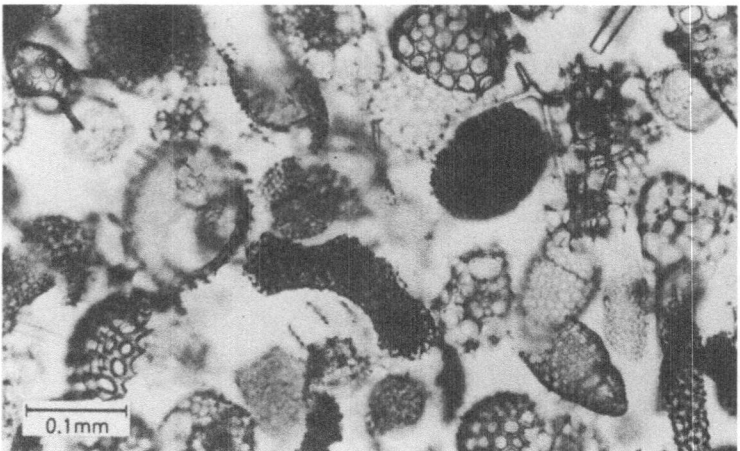

Fig. 13. Radiolarians and radiolarian fragments separated from the sediment from radiolarian ooze belt.
Photo: H. Schwetje BGR. Magnification: 200 ×

pitation, they are forming a manganese nodule around it. A fraction of the iron and manganese hydroxides precipitated within the sediment as micronodules (Figs. 14–16) (with dimensions ranging from <1 mm to ~2 mm) which generally contain a radiolarian or a particle of volcanic ash as a nucleus. The rest of iron and manganese really constituent part, is precipitated in a disseminated form within the sediment.

All these primary precipitates contain more iron than manganese, as they precipitate directly from the sea-water which is enriched in iron compared to manganese.

The pore waters within the first few decimeters of sediments are even more enriched in metals, needed for the precipitation of manganese nodules, than the bottom-near sea-water.

These metals (Mn, Cu, Ni, Zn) are evidently mobilized from the sediment. There are several different theories about the cause of this mobilization.

According to our investigations the organic carbon content decreases with depth in the sediment. This may be caused by changes in the primary production, but more probably by the oxidation of organic carbon within the sediment. This oxidation would lower the redox potential so that manganese would be dissolved. Our measurements of the redox potential in the sediment samples on board ship confirm this theory, but these measurements are never reliable enough, because the sediment was transported about 2 hours from the sea bottom to the surface. During that time the pressure decreased from 500 to 1 bar and the temperature increased. All these factors may cause the change of the redox potential.

We have also observed the partial dissolution of micronodules within the sediment. The elements which are first dissolved from micronodules are primarily the

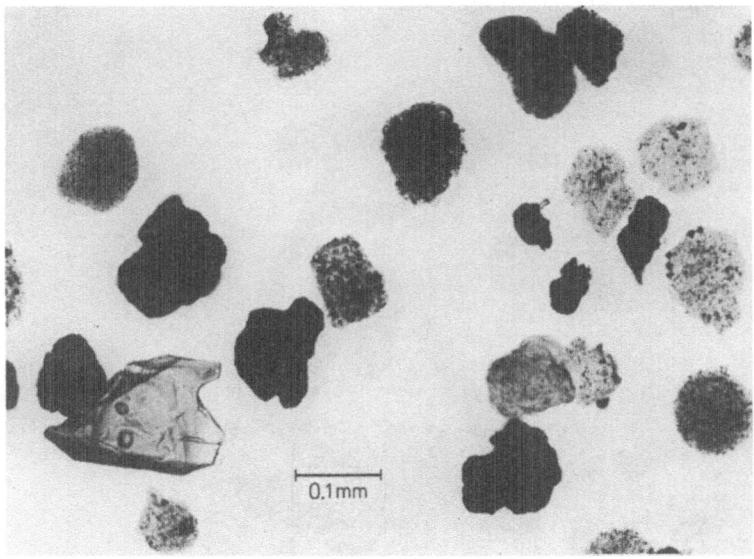

Fig. 14. Different particles occurring in radiolarian ooze Black: micronodules, patchy: clay aggregates, glassy sharp-edged fragment: volcanic ash. (Clay aggregates can partly be formed as weathering products of volcanic ash or partly be new built).
Photo: H. Schwetje BGR. Magnification: 200 ×

119

Fig. 15. Rounded manganese micronodules separated magnetically from radiolarian ooze. They usually overgrow radiolarians or clay mineral aggregates.
Photo: E. Rehm, TU Clausthal. Magnification 50 ×

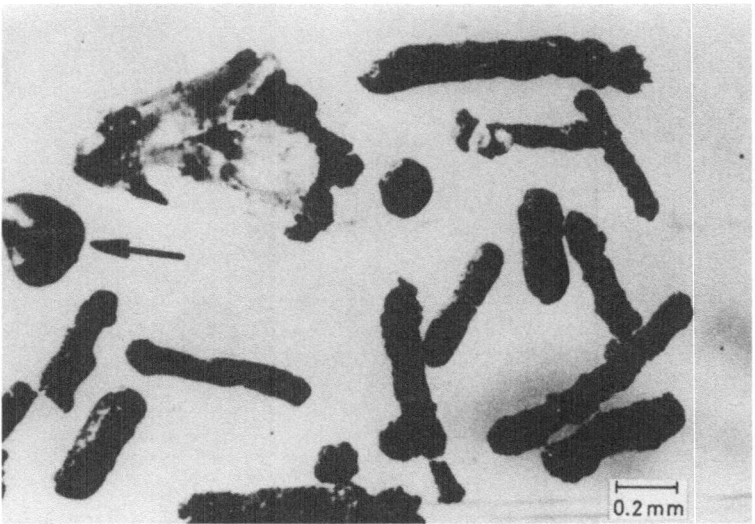

Fig. 16. Elongated shape of manganese micronodules (due to filling of small worm holes or overgrowing sponge needles).
Photo: E. Rehm, TU Clausthal. Magnification 50 ×

120

most mobile ones: Mn, Ni, Cu, and Zn. As dissolution proceeds the micronodules become enriched in Fe, Si, and Al in comparison to Mn, Ni, Cu, and Zn [22].

Hartmann[12] observed that the worm traces are depleted in Mn, Ni, Cu, and Zn in comparison to the surrounding sediment. His opinion is that the decomposition of organic matter cannot lower the redox potential so far that dissolution of manganese takes place, except in locally limited patches where organic matter is higher because of bioturbation.

Schellmann (unpublished results) calculated from the solubility products of manganese and iron hydroxides that interstitial water can be sufficiently supplied with manganese and iron even at higher redox potentials, because the time of dissolution is adequately long.

Part of the material used for the enrichment of pore water in the elements mentioned seems to come from the dissolution of radiolarians. They absorb especially copper and zinc from the sea-water during the sinking to the sea bottom[11]. As the radiolarians are dissolved to 50% within the first 35 cm of the sediment column, Cu and Zn absorbed on their surfaces are dissolved.

All the former authors confirm that a remobilization of metals within the sediment takes place. As the sediment is very porous because of its content of radiolarians, the dissolved material can diffuse upwards in the direction of smaller concentrations in the sea-water. These authors also confirm that the quantity of the mobilized metals within the sediment is higher than the need for the supply of manganese nodules during their growth.

The mobilized metals are reprecipitated in the nearly liquid layer on the sediment water interface. They reprecipitate partly on the surface of manganese nodules located there. The other part precipitates within the sediment, forming again micronodules or disseminated precipitates. A portion of the metals again diffuses into bottom-near sea-water.

Characteristic of the precipitates of remobilized material dissolved in pore water is that the manganese in this material strongly predominates over iron.

In this way, manganese nodules growing on the radiolarian ooze belt are supplied from two sources with metals needed for their growth: from the sea-water with iron-rich material and from the pore-water with manganese-rich material.

The nodules growing on the top of the seamounts in the same area can get their supply only from the sea water, because of the lack of sediment. Consequently, their chemical composition shows the predominance of iron.

The supply of manganese nodules, growing on the sediment, with iron-rich precipitates from the sea-water and manganese-rich precipitates from the pore-water also results in some differences of the chemical composition: the sediment side of the nodule is richer in manganese and the water side of the same nodule is richer in iron.

As manganese nodules receive a portion of the metals for supply of their growth from the sediment, it may be expected that the sedimentation rate must have an influence on the growth of manganese nodules and their chemical composition. Slow sedimentation rates would be optimal because the diagenetic dissolution would have enough time to take place. The interuption of sedimentation would stop the supply of nodules because, after a while, the sources of remobilization in the sediment would be exhausted. Too quick sedimentation would cover the nodules;

and prevent them from staying on the sediment-water interface, which is the optimal medium for their growth (see Fig. 10 and 11).

Price and Calvert [29] could confirm this theory with the observation that Mn/Fe ratios in the manganese nodules are negatively correlated with the accumulation rates of the sediment. The slower the sedimentation, the more manganese-rich remobilized metal solution from the sediment contributes to the growth of the nodules.

If manganese nodules are cut and polished (Fig. 8) it can be observed that they are built of a great number of concentric layers around the nucleus. Some of these concentric layers are laminated, and some of them show the structures of dendritic growth. The laminated layers were grown more slowly and the dendritic layers more quickly [13]. The laminated layers are also richer in iron whereas the dendritic layers are richer in manganese [16].

From this it can be concluded that during the growth of dendritic layers the conditions for the remobilization of manganese from the sediment are better so that manganese-rich pore water can contribute a great deal to the growth of the layer. During the growth of a laminated layer the supply from the sediment is small or stopped at all so that the layer is built mostly of the precipitation from iron-rich sea-water.

In good agreement with these observations is the fact that manganese nodules growing on the deep-sea hills (i.e. without supply from the sediment) are built only of laminated layers. The dendritic layers are totally missing in these nodules (Fig. 9).

It can be calculated for the investigated area that the contribution of material from the sediment is three times higher than that from the sea-water [14]. This accounts for the great amount of manganese-rich nodules growing on the sediment and the small amount of iron-rich nodules growing on places without sediment cover, e.g. deep-sea hills.

This contribution from the sediment obviously changed several times within the time of the growth of the nodule; therefore, we can find manganese-rich (dendritic) layers and iron-rich (laminated) layers in turns within one nodule from deep-sea plains.

The deep-sea radiolarian ooze is a biologically active zone as we could observe by the bioturbation of the sediments. Thus, the question arises how far the biologic processes can influence the growth of manganese nodules.

On the surface of manganese nodules also some life could be observed: small pipes of agglutinating foraminifera and fungi mycels overgrowing the nodules like spider web are the most frequent forms of life.

Manganese-oxidizing bacteria could also be identified [32]. As manganese in the sediments has to be reduced to Mn^{2+} ions to become soluble and thus to get mobilized, the very important question arises how or where it is reoxidized. Generally, the bottom-near sea-water of Central Pacific is so rich in dissolved oxygen that manganese and iron will be oxidized when coming into the contact with sea-water (at the sediment-water interface and in the mobile, nearly liquid uppermost sediment layer).

Manganese-oxidizing bacteria living on the sediment-water interface do not change the chemical processes which could also take place without them. They only promote the initiation of these processes and may accelerate them. Thus, they probably form a kind of a "catalyst".

With the description of the special case of the genesis of manganese nodules I hope to have explained how complex the genesis of this type of nodule can be. We are far away from the possibility of giving a detailed description of the genesis of any other type of manganese nodule field.

9 Economic and Legal Problems Connected with the Mining of Manganese Nodules

Within the last twenty years deep-sea manganese nodules (which have already been known for about one hundred years) have come to discussion as a multi-element ore of the future. The interesting metals in this ore are copper, nickel, cobalt, and manganese. Zinc and molybdenum could also be used as by-products.

The best mining area known so far is the radiolarian ooze belt in the Central Pacific. Pearson[28] reported the following average composition of nodules for this area:

1.16% Cu	0.14% Zn
1.28% Ni	24.6 % Mn
0.23% Co	0.06% Mo .

These values are very similar to the average we obtained from chemical analyses of our nodules from the plains. Mero 1977 calculated for this area (6,000,000 km^2) an average nodule density of 9 kg/m^2, i.e. 54 billion tons of wet nodules or 38 billion tons of dry nodules.[25]

The whole Pacific Ocean floor contains 1.5 trillion tons of manganese nodules[28] but the percentages of commercially useful metals are lower for the average of the total Pacific than in the radiolarian ooze area.

The knowledge of most areas is too scarce up to now to make really good estimations of the manganese nodule content and its chemistry. A great deal of sampling and extensive analyses are being performed by industrial research being however not available for publications.

The manganese nodules of the described quality and quantity would be an extremely rich ore if positioned on the earth surface. The procedure of deep sea mining and transport will raise the costs so much that this mining will be unprofitable now.

There are several factors which can promote the profitability of manganese nodule mining:

a) The coverage of nodules should be high enough so that the mining vessel has to go shorter distances by mining larger quantities,
b) the content of economically important metals in manganese nodules should be high,
c) manganese nodules should lie on the soft sediment and not on the rocky grounds,

d) the topography of the field should be as smooth as possible,
e) an important factor is also the low percentage of "gangue material" i.e. stones and fossil fragments which are, because of their dimensions, mined together with manganese nodules,
f) the weather in the mining area should be calm enough to make mining possible for the larger part of the year.

Additionally, the water depth of the mining area and the distance to the metallurgic plant should be taken into account in profitability calculations.

Mero calculated[25] that a manganese nodule field with more than 2.8 % Ni + Cu + Co in dry manganese nodule matter, with nodule concentrations higher than 5 kg/m^2 and with less than 20 % of ballast material (in an area which permits mining in more than 250 days of the year), may be considered to be of economic profitability.

These data will probably change from year to year according to the varying prices of metals and the developing technology of deep-sea mining.

Of interest is also the following calculation[26]: if 1 % of the supposed reserves of manganese nodules was mined, then this mining would supply the world market with quantities of nickel and manganese as high as the complete terrestrial reserves of these metals, the quantity of copper corresponding to one tenth of the terrestrial reserves and the quantity of cobalt being twelve times as high as the terrestrial reserves. Knowing this, it is imaginable that mining of nodules would strongly influence the world metal market. Especially in the case of cobalt the prices would fall considerably which would again change the profitability calculations for deep-sea mining.

The major part of economically important metals from manganese nodules are alloys for steel production (Mn, Ni, Co, Mo). The production of steel is supposed to grow continuously in the next decades, because of the industrialization of underdeveloped countries. Thus, it can be expected that the prices of these metals will rise continuously, and the time will come when manganese nodules will be able to compete on the world trade market as a multi-element ore. Mez[26] means it will be about the year 2000.

In the near future the development of mining technologies seems to concentrate on hydraulic sucking systems. Pilot tests are being made.

In the development of metallurgic processes for obtaining different metals from the multimetal ore manganese nodules, the gaps in the knowledge concerning the type of bonding of the elements in the manganese nodules are a great handicap.

In spite of that different leaching processes were developed and tested in the laboratories including acid dissolution with subsequent separation of metals, selective acid leaching for Cu, Ni and Co, ammoniacal leaching etc.

The pilot tests and the demand for metals will decide which of the proposed leaching methods will be used.

The legal problems of deep-sea mining are not yet solved because the zones of nodule mining will usually be out of the 200-mile national zones.

There still holds a medieval international principle about the liberty of the sea. Anybody possessing the technology needed may raise nodules from the sea floor.

The "Moratorium" resolution of 1969 and the UN resolution of 1970 proclaim the principle of the sea as a heritage of the whole human race. Thus the nations having the technological possibilities should exploit the seas and share the profit with underdeveloped nations.

Since the UN resolutions are only recommendations and not international law they do not have to be respected by the nations or companies intending to mine manganese nodules.

The 3rd UN conference about the sea-law, aiming at solving the problems of exploiting the deep sea was organized in 1974. It is still working and it dos not seem, that any decisions will be reached within the next few years.

10 References

1. Beiersdorf, H.: Ergebnisse der Manganknollen-Wissenschaftsfahrt VA 13/1, fachlicher Bericht, BGR Hannover (1976)
2. Bischoff, J. L., Dickson, F. W.: Earth Planet. Sci. Lett. *25*, 385 (1975)
3. Burns, R. G.: Nature *205*, 999 (1965)
4. Burns, R. G., Burns, V. M.: Phil. Trans. Roy. Soc. (London) *A 285*, 249 (1977)
5. Cronan, D. S.: Underwater minerals. A.P. 1980
6. Dreyfus Rawson, M., Ryan, W. B. F.: Ocean floor sediment and polymetallic nodules, Columbia Univers., New York, 1978
7. Giovanoli, R., Brütsch, R.: Chimia *33*, 372 (1979)
8. Glasby, G. P.: Marine Geol. *28*, 51 (1978)
9. Gundlach, H., Marchig, V., Schnier, C.: Geol. Jb. *D 23*, 67 (1977)
10. Hain, J. R., et al.: In: Marine geology and oceanography of the Pacific manganese nodule province. Bischoff, J. L., Piper, D. C. (eds.), P. 365. Plenum Press, New York, 1979
11. Halbach, P., Rehm, E., Marchig, V.: Marine Geol. *29*, 237 (1979)
12. Hartmann, M.: Chem. Geol. *26*, 277 (1979)
13. Heye, D.: Geol. Jb. *E 5*, 3 (1975)
14. Heye, D.: Marine Geol. *28*, M59 (1978)
15. Heye, D.: Proc. Colloques Internationaux du CNRS no. 289. La genèse des nodules de manganèse, Paris, 1979
16. Heye, D., Marchig, V.: Marine Geol. *23*, M19 (1977)
17. Lalou, C., Brichet, E.: Mineral. deposita *11*. 267 (1976)
18. Lalou, C., et al.: Proc. Colloques Internationaux du CNRS no. 289. La genèse des nodules de manganèse, Paris 1979
19. Lalou, C., Brichet, E., Jehanno, C.: Proc. Colloques Internationaux du CNRS no. 289. La genèse des nodules de manganèse, Paris, 1979
20. Lisitzin, A. P.: Sedimentation in the world ocean. Soc. Econ. paleont. miner., Spec. Publ. No. 17, 1972
21. Livingstone, D.: US Geol. Surv. profess. paper *440G* (1963)
22. Marchig, V., Gundlach, H.: Proc. Colloques internationaux du CNRS, no. 289. La genèse des nodules de manganèse, Paris, 1979
23. Marchig, V., Gundlach, H., Schnier, C.: Geol. Rdsch., *68*, 1037 (1979)
24. Mero, J. L.: The mineral resources of the sea. Amsterdam: Elsevier 1965
25. Mero, J. L. In: Marine manganese deposits. (Glasby, G. P., ed.), p. 327. Elsevier, Amsterdam, 1977
26. Mez, B.: Marine Rohstoffgewinnung. Proc. 7, Seminar Meerestechnik TU Clausthal/TU Berlin, 1979
27. Mottl, M. J., Holland, H. D., Corr, R. F.: Geochim. Cosmochim. Acta *43*, 869 (1979)
28. Pearson, J. S.: Ocean floor mining. Ocean technology review No. 2, 1975
29. Price, N. B., Calvert, S. E.: Marine Geol. *9*, 145 (1970)
30. Rösler, H. J., Lange, H.: Geochemische Tabellen, Leipzig 1975

31. Schnier, C., Gundlach, H., Marchig, V.: Proc. Third International Symposium on Environmental Biogeochemistry, vol. 3, p. 859, Wolfenbüttel, 1977
32. Schütt, C., Ottow, J. C. G.: Z. allg. Mikrobiol. *17*, 8, 611 (1977)
33. Seibold, E.: Geol. Rdsch. *62*, 641 (1973)
34. v. Stackelberg, U. In: Marine geology and oceanography of the Pacific manganese nodule province (Bischoff, J. L., Piper, D. Z., ed.), p. 559. Plenum Press, New York, 1979
35. Turekian, K. K., Scott, M. R.: Environ. Sci. Technol. *1*, 940 (1967)
36. Wedepohl, K. H.: 10th Int. Geol. Congress, Sydney 1976. In: International monograph on geology and geochemistry of manganese. Varentsov, I. M. (ed.). Budapest: Publishing House of the Hungarian Academy of Sciences 1979 (in press)
37. Zuleta Roncal, R.: Geochemische Untersuchungen am Meerwasser und Manganknollen im Bereich des Zentralen Pazifischen Ozeans. Diss. TH Aachen 1976

Author Index Volumes 50–99

Dürr, H.: Triplet-Intermediates from Diazo-Compounds (Carbenes). *55*, 87–135 (1975).
Dürr, H., and Kober, H.: Triplet States from Azides. *66*, 89–114 (1976).
Dürr, H., and Ruge, B.: Triplet States from Azo Compounds. *66*, 53–87 (1976).
Dugundji, J., Kopp, R., Marquarding, D., and Ugi, I.: A Quantitative Measure of Chemical Chirality and Its Application to Asymmetric Synthesis *75*, 165–180 (1978).
Dumas, J.-M., see Trudeau, G.: *93*, 91–125 (1980).
Dupuis, P., see Trudeau, G.: *93*, 91–125 (1980).

Eicher, T., and Weber, J. L.: Structure and Reactivity of Cyclopropenones and Triafulvenes. *57*, 1–109 (1975).
Eicke, H.-F., Surfactants in Nonpolar Solvents. Aggregation and Micellization. *87*, 85–145 (1980).
Epiotis, N. D., Cherry, W. R., Shaik, S., Yates, R. L., and Bernardi, F.: Structural Theory of Organic Chemistry. *70*, 1–242 (1977).
Eujen, R., see Bürger, H.: *50*, 1–41 (1974).

Fischer, G.: Spectroscopic Implications of Line Broadening in Large Molecules. *66*, 115–147 (1976).
Flygare, W. H., see Sutter, D. H.: *63*, 89–196 (1976).
Frei, H., Bauder, A., and Günthard, H.: The Isometric Group of Nonrigid Molecules. *81*. 1–98 (1979).

Gandolfi, M. T., see Balzani, V.: *75*, 1–64 (1978).
Ganter, C.: Dihetero-tricycloadecanes. *67*, 15–106 (1976).
Gasteiger, J., and Jochum. C.: EROS — A Computer Program for Generating Sequences of Reactions. *74*, 93–126 (1978).
Geick, R.: IR Fourier Transform Spectroscopy. *58*, 73–186 (1975).
Geick, R.: Fourier Transform Nuclear Magnetic Resonance, *95*, 89–130 (1981).
Gerischer, H., and Willig, F.: Reaction of Excited Dye Molecules at Electrodes. *61*, 31–84 (1976).
Gleiter, R], and Gygax, R.: No-Bond-Resonance Compounds, Structure, Bonding and Properties. *63*, 49–88 (1976).
Gleiter, R. and Spanget-Larsen, J.: Some Aspects of the Photoelectron Spectroscopy of Organic Sulfur Compounds. *86*, 139–195 (1979).
Gleiter, R.: Photoelectron Spectra and Bonding in Small Ring Hydrocarbons. *86*, 197–285 (1979).
Gruen, D. M., Veprek, S., and Wright, R. B.: Plasma-Materials Interactions and Impurity Control in Magnetically Confined Thermonuclear Fusion Machines. *89*, 45–105 (1980).
Guérin, M., see Trudeau, G.: *93*, 91–125 (1980).
Günthard, H., see Frei, H.: *81*, 1–98 (1979).
Gygax, R., see Gleiter, R.: *63*, 49–88 (1976).

Haaland, A.: Organometallic Compounds Studied by Gas-Phase Electron Diffraction. *53*, 1–23 (1974).
Hahn, F. E.: Modes of Action of Antimicrobial Agents. *72*, 1–19 (1977).
Hargittai, I.: Gas Electron Diffraction: A Tool of Structural Chemistry in Perspectives, *96*, 43–78 (1981).
Hayatsu, R., see Anders, E.: *99*, 1–37 (1981).
Heaton, B. T., see Chini, P.: *71*, 1–70 (1977).
Heimbach, P., and Schenkluhn, H.: Controlling Factors in Homogeneous Transition-Metal Catalysis. *92*, 45–107 (1980).
Hendrickson, J. B.: A General Protocol for Systematic Synthesis Design. *62*, 49–172 (1976).
Hengge, E.: Properties and Preparations of Si-Si Linkages. *51*, 1–127 (1974).
Henrici-Olivé, G., and Olivé, S.: Olefin Insertion in Transition Metal Catalysis. *67*, 107–127 (1976).
Hobza, P. and Zahradnik, R.: Molecular Orbirals, Physical Properties, Thermodynamics of Formation and Reactivity. *93*, 53–90 (1980).
Höfler, F.: The Chemistry of Silicon-Transition-Metal Compounds. *50*, 129–165 (1974).
Hogeveen, H., and van Kruchten, E. M. G. A.: Wagner-Meerwein Rearrangements in Long-lived Polymethyl Substituted Bicyclo[3.2.0]heptadienyl Cations. *80*, 89–124 (1979).
Hohner, G., see Vögtle, F.: *74*, 1–29 (1978).

Cosmochemistry

1974. 48 figures. IV, 176 pages
(Topics in Current Chemistry, Volume 44)
ISBN 3-540-06457-5

Contents/Information:

G. Winnewisser, P. G. Mezger, H.-D. Breuer:
Interstellar Molecules
This is a review of organic and inorganic
interstellar matter and refers to observations
of molecules in circumstellar shells and
molecule formation in protostellar nebulae.
(76 references)

G. Eglinton, J. R. Maxwell, C. T. Pillinger
**Carbon Chemistry of the Apollo Lunar
Samples**
There is a small magmatic component of
methane in some crystalline rocks but the
gaseous hydrocarbons in the fines derive
mainly from solar-wind implantation. The
implantation effects occurring on the lunar
surface also have general implications for
the synthesis of the molecules detected in
interplanetary and interstellar dust.
(118 references)

H. Wänke
Chemistry of the Moon
The moon is a highly differentiated object.
Many chemical elements on the moon are
strongly enriched or depleted as compared
with their abundance in carbonaceous chondrites which, apart from the most volatile
elements, are believed to be representative
of solar matter. (132 references)

Springer-Verlag
Berlin
Heidelberg
New York

Inorganic Chemistry Concepts

Editors: M. Becke, C. K. Jørgensen, M. F. Lappert,
S. J. Lippard, J. L. Margrave, K. Niedenzu,
R. W. Parry, H. Yamatera

Springer-Verlag
Berlin Heidelberg New York